AI Powered Project Management

The Future of Strategic Execution

By Lynn R. Squire

Disclaimer

Contents

Chapter 1: Introduction to AI in Project Management

Project management isn't always a walk in the park, but with a well-oiled team, you can achieve almost anything. Still, the problem lies in the mundane and repetitive tasks that can make any team member lose interest or slow progress. Of course, humans will get tired – they need rest. But what if there were a solution to this problem? What if you could have an assistant who never sleeps, doesn't complain about dull everyday tasks, and works all the time to complete projects effortlessly? That's the promise of artificial intelligence (AI) in project management.

AI is no longer just a buzzword; it has become a game-changer – that is if it is used correctly. With the right command over AI, you can make project management a breeze. You get to control how projects are planned, executed, and delivered. The best part is that AI is not reserved for tech wizards or massive corporations with endless budgets. AI is for everyone, and its role in project management is simple but powerful: *to handle the routine so you can focus on the extraordinary.*

How? Let me explain.

The most tedious parts of your workflow can be automated, content can be generated within seconds, and your schedules can be optimized without you even lifting a finger. When AI handles all these repetitive tasks for you, you end up with more free time that can be used to make strategic decisions that drive real progress.

AI-Powered Project Management is here to make that transformation accessible at any stage. In the pages to follow, you will find practical advice and knowledge that you can apply immediately. Each chapter is packed with resources to help you learn and implement AI in ways that matter most to your projects.

Of course, this doesn't mean AI will replace your expertise – it will simply enhance it. AI clears the clutter so you can excel at what humans do best: thinking creatively, solving problems, and leading with purpose. While the idea of adopting AI might feel discouraging to some, you'll soon see how intuitive and empowering these tools can be.

The Current State of AI in Project Management

AI in project management remains largely unexplored. According to a recent survey, only 12% of project managers have made substantial use of AI.[1] The problem is that most people do not fully comprehend how AI can empower companies and organizations in project management and goal achievement.

Many project managers still rely on traditional tools and techniques, like spreadsheets or standalone project management software, to coordinate tasks and track progress. While these tools have their merits, they often fall short when dealing with complex projects that require agility, real-time updates, and predictive insights. AI has the potential to bridge this gap, but adoption has been slow due to a lack of awareness, resistance to change, and misconceptions about its complexity.

---

[1]https://rebelsguidetopm.com/ai-in-project-management-statistics/

AI is gradually making inroads in areas like task automation, predictive analytics, and resource optimization. For example, AI-powered tools can analyze historical project data to predict potential risks or delays, helping managers take proactive measures. Automated scheduling systems can allocate resources more effectively, while natural language processing (NLP) tools can streamline communication by generating summaries of meetings or automatically tagging relevant team members in updates.

Despite these advancements, the integration of AI in project management is far from mainstream. A significant challenge lies in the perception that AI is a luxury reserved for large, tech-savvy organizations. Small and medium-sized enterprises (SMEs), which form a major chunk of the global economy, feel AI is beyond their reach due to cost or technical barriers.

The reality is AI is becoming more accessible than ever. Many affordable AI-driven project management tools are now on the market, designed to cater to teams of all sizes. Platforms like Monday.com, Asana, and Trello are already incorporating AI features, such as task prioritization and workload balancing, making it easier for teams to dip their toes into the world of AI without overhauling their entire workflow.

Another issue holding back AI adoption is the fear of job displacement. 1 in 3 workers in the US has concerns about job displacement due to AI use.[2] The truth is that AI is not replacing people; rather, it is augmenting their abilities. Ultimately, it allows project managers to focus on strategy,

_____

[2] https://www.aiprm.com/ai-replacing-jobs-statistics/

leadership, and creative problem-solving instead of working on redundant and time-consuming operations.

The current state of AI in project management can best be described as a tipping point. The tools are available, the benefits are clear, and early adopters are already reaping the rewards. What's needed now is a shift in mindset. As more organizations see AI not as a futuristic luxury but as a practical solution to real-world challenges, its adoption will accelerate.

Understanding the Potential and Limitations of AI in Project Management

AI is a powerful tool that can enhance productivity, efficiency, and decision-making in project management. However, it is not a replacement for human expertise.

Success lies in using AI as a partner to augment human capabilities. But to do that, you must first familiarize yourself with the potential and limitations of AI. Let's take a look:

Potential

AI possesses the ability to reshape project management. It offers tools and insights that can revolutionize how teams work.

To start with, AI excels at processing vast amounts of data quickly and accurately.

It analyzes historical project data to identify patterns, forecast potential delays, and suggest solutions to prevent possible setbacks. Predictive analytics allows teams to anticipate risks and opportunities, which enables proactive planning with minimal or no disruptions.

Other areas of AI potential in project management include:

Automation of Mundane Tasks
Repetitive tasks such as scheduling, sending reminders, or updating project timelines are not only time-consuming but also drain productivity. However, AI-powered tools handle these tasks effortlessly to free up time for more meaningful work.

For instance, AI scheduling tools can dynamically adjust timelines based on changes in project scope or resource availability.

Enhanced Communication and Collaboration
AI-driven chatbots and virtual assistants streamline the communication process by providing instant updates, answering routine queries, and keeping team members informed. These tools reduce unnecessary back-and-forth emails or meetings, which significantly improves the overall efficiency of the team.

Resource Optimization
Allocating resources effectively is a challenge for any project manager. But AI can make the process much simpler.

It can analyze every team member's skills, workload, and availability to recommend the best allocation of resources to ensure no one is overburdened or underutilized.

In a way, it creates a balance, which ultimately promotes better performance and team satisfaction.

Real-Time Monitoring and Adjustments
AI tools provide real-time updates on project progress, budgets, and timelines. Managers can quickly spot deviations and make informed decisions to get projects back

on track. As a result, all projects stay in line with the final goals while effortlessly meeting deadlines.

Limitations

Now, let's explore the limitations of AI in project management:

Dependency on Data Quality

AI is only as good as the data it processes. Poor-quality, incomplete, or biased data means inaccurate predictions and flawed recommendations.

It is important to always provide clean, accurate, and comprehensive data for AI to function effectively.

Lack of Context and Emotional Intelligence

AI is great at identifying trends and patterns, but it lacks the ability to understand the context or nuances behind them. It cannot assess interpersonal dynamics, manage conflicts, or address ethical dilemmas like a human could. These tasks require human empathy, judgment, and communication skills, which AI cannot replicate.

Challenges in Adoption

For many organizations, integrating AI into existing workflows is a daunting task. The learning curve for new tools, resistance to change, and limited technical expertise among team members are slowing the pace of adoption.

In many smaller organizations, budget constraints are another issue, which makes it harder to justify the initial investment in AI technology.

Limitations in Creativity and Innovation

AI works based on patterns and historical data, but it cannot innovate or think creatively. Of course, it can suggest solutions to known problems; however, it won't come up with groundbreaking or out-of-the-box ideas or adapt to unique scenarios that fall outside its programmed algorithms.

Ethical Considerations and Responsible AI Use

With the growing popularity of AI in project management, it is crucial to factor in responsible usage and ethical considerations to maintain trust, fairness, and accountability.

How exactly can you do that? Here are a few factors to consider:

Transparency and Accountability

AI systems sometimes operate like a "black box." They make decisions based on algorithms that users don't fully understand. When these decisions influence critical aspects of a project – such as resource allocation or risk predictions – team members need to know how the AI arrived at its conclusions. Transparency is key here.

Project managers should use AI tools or models that provide explainable insights. It ensures stakeholders can trust the process and outcomes.

At the same time, accountability holds equal importance. If an AI system makes a mistake, who is responsible? Organizations should always define accountability structures.

In the end, project managers and team members are responsible for the decisions they make. It is important to understand that AI is simply a tool to support. It does not replace human judgment.

Bias and Fairness

AI systems learn from data, and if the input data contains biases, the AI's output will reflect them. For example, if historical data used for resource allocation favors certain team members over others, AI might perpetuate or even amplify that bias.

Responsible AI use requires vigilance in identifying and mitigating such biases while training data and decision-making processes. It is essential that teams regularly audit AI systems for fairness. This ensures that outputs streamline with ethical standards and do not disadvantage any group or individual unfairly.

Privacy and Data Security

AI tools sometimes require access to sensitive data. Protecting this data is critical.

Unauthorized access or misuse may lead to breaches of confidentiality, reputational damage, or even legal consequences. That's why organizations must implement strong data protection measures, including encryption, secure storage, and strict access controls.

Additionally, teams should comply with relevant privacy regulations, such as GDPR, to safeguard individual and organizational data rights.

Chapter 2: Foundational AI Technologies for Project Management

According to an early study by Gartner, AI will execute almost 80% of project management tasks by 2030.[3]

In theory, the processes will become more streamlined, efficiency will be higher than ever, and decision-making will be the best there is. Sounds too good to be true, right? But it can and will make all of this happen. The key is to pick the right technologies.

That said, let's take a look at some of the foundational AI technologies that can be used in project management.

Large Language Models (LLMs)

Large Language Models (LLMs) are a type of artificial intelligence designed to understand, generate, and analyze human language. They are built using advanced machine learning techniques and trained on massive datasets that contain text from books, articles, websites, and other written content. These models can comprehend context, detect intricacies in language, and generate coherent, human-like text based on the input they receive.

Neural networks form the backbone of LLMs. In most cases, they use transformer architecture. Transformer architecture is a sequence-to-sequence model that finds meaning for input data using neural networks. It enablesLLMs to process text efficiently, analyze relationships between words, and predict the most likely next word or phrase in a sentence.

[3]https://www.gartner.com/en/newsroom/press-releases/2019-03-20-gartner-says-80-percent-of-today-s-project-management

Over time, this process builds a rich understanding of language.

Applications of LLMs in Project Management

LLMs have the potential to save time, improve communication, and optimize workflows. Some practical uses include:

- **Drafting and Editing Documents:** LLMs can write project reports, meeting minutes, and email updates, among other things. They ensure your content is clear, concise, and error-free.
- **Generating Ideas:** Brainstorming sessions can always be enhanced by using LLMs. You can generate creative solutions or alternative approaches at any given time.
- **Analyzing Text Data:** Feedback from stakeholders or survey responses can be analyzed to identify patterns and areas for improvement.
- **Summarizing Information:** Long documents, such as research reports or project proposals, can be condensed into key points to make them easier to understand.
- **Automating Routine Queries:** LLMs can answer common questions regarding project details, deadlines, or progress, which can free up a lot of time for team members.

Advantages of LLMs in Project Management

The integration of LLMs into project management brings several advantages:

- **Time-Saving:** Automation of repetitive tasks like drafting emails or summarizing documents allows teams to focus on more strategic activities.

- **Improved Communication:** LLMs ensure consistency and professionalism in written communication, reducing the chances of possible misunderstandings.
- **Scalability:** When LLMs are handling large volumes of text data, such as client feedback or status updates, the workload becomes much more manageable.
- **Cost Efficiency:** Automating tasks typically handled by humans reduces operational costs in the long run.
- **Enhanced Decision-Making:** Insights generated by LLMs from analyzing data help project managers make more informed decisions faster.

Popular LLMs for Project Management

GPT (Generative Pre-trained Transformer)
Developed by OpenAI, GPT is an advanced AI language model. It's built to interpret and produce human-like text based on the input it receives.

GPT can also analyze vast amounts of text data and produce coherent and contextually relevant responses.

Key Features of GPT
- **Natural Language Understanding and Generation:** GPT excels at comprehending and generating human-like text, which allows for seamless interactions and content creation.
- **Contextual Comprehension:** It can maintain context over extended conversations or text. As a result, all the responses remain relevant and coherent.
- **Adaptability:** GPT can be fine-tuned for specific tasks based on the industry. This enhances its effectiveness in specialized applications.

- **Multilingual Support:**It supports multiple languages, which means it facilitates communication across diverse teams and stakeholders.

Applications in Project Management

- **Automating Routine Tasks:**GPT can draft meeting agendas, project charters, and risk registers, thus reducing the administrative burden on project managers.
- **Summarizing Information:** It can condense lengthy documents or discussions into concise summaries, which helps in making quick decisions.
- **Generating Ideas:** During brainstorming sessions, GPT can provide creative suggestions or alternative approaches to overcome project challenges.
- **Analyzing Text Data:** It can process feedback from stakeholders or team members, which helps point out common themes and areas for improvement.
- **Real-Time Assistance:** GPT can answer queries about project details, timelines, or best practices on the spot, playing the role of a virtual assistant.

Pricing

OpenAI offers various pricing tiers for GPT access:

- **Free Access:** The best thing about GPT's is that it can be used at no cost. Of course, the free model comes with limitations on features and usage.
- **ChatGPT Plus:** For $20 per month, this plan provides general access even during peak times, faster response times, and priority access to new features and improvements.
- **ChatGPT Pro:** The Pro model is aimed at heavy users and professionals. It has a $200 monthly

subscription, which offers unlimited access to advanced models like GPT-4o and o1, along with exclusive features such as o1 pro mode and Advanced Voice.

There are also **Team** and **Enterprise** models for GPT with different pricing plans.

Gemini

Gemini is an advanced AI system developed by Google DeepMind. It is designed to process and understand various forms of data.

Its multimodal capabilities enable it to perform complex tasks, making it a versatile tool for numerous applications.

Key Features of Gemini

- **Multimodal Processing:** Gemini can seamlessly interpret and generate content across different data types, such as text, images, audio, and video, allowing for more comprehensive and integrated outputs.
- **Advanced Problem-Solving:** Gemini is equipped with sophisticated reasoning abilities, which allows it to tackle complex challenges and provide insightful solutions and recommendations.
- **Creative Content Generation:** Gemini excels in producing creative content, including writing, image creation, and multimedia presentations, enhancing the quality and engagement of project deliverables.
- **Integration with Google Workspace:** Gemini can be integrated with Google Workspace to enhance productivity tools like Docs, Sheets, and Slides with AI-driven features.

Applications in Project Management

- **Resource Allocation:** Gemini analyzes project requirements and assists in distributing resources effectively to ensure optimal utilization.
- **Timeline Management:** It helps create detailed project timelines with milestones, which makes assigning tasks to team members and stakeholders much more efficient.
- **Data Analysis:** Gemini can process large datasets to extract meaningful insights. This, in turn, aids in informed decision-making throughout the project lifecycle.
- **Communication Enhancement:** It generates clear and concise reports, summaries, and presentations, which is extremely beneficial for improving communication among team members.

Pricing

Gemini offers a range of subscription options to cater to diverse user needs:

- **Gemini API Access:** The Gemini API model can be utilized through Google AI Studio. It offers a free tier suitable for testing purposes. The free tier provides 15 requests per minute (RPM) and 1 million tokens per minute (TPM). For more extensive usage, a pay-as-you-go pricing model is available, featuring higher rate limits and costs based on token consumption. For instance, input pricing starts at $0.075 per 1 million tokens, and output pricing at $0.30 per 1 million tokens.
- **Gemini Advanced Subscription:** The advanced subscription plan is priced at $19.99 per month. It includes a two-month free trial, allowing users to

explore its features before committing. Subscribers also gain access to the Ultra-powered chatbot, enhanced multimodal capabilities, and integrations with Google apps like Gmail and Docs. Additionally, they receive 2 TB of Google One storage and priority access to new features.

LLaMA (Large Language Model Meta AI)

LLaMA, developed by Meta (formerly Facebook), is an advanced AI language model designed to offer cutting-edge generative AI capabilities. With a focus on efficiency, scalability, and adaptability, LLaMA brings powerful tools to streamline workflows and enhance productivity in project management. Its unique design ensures accessibility for teams of all sizes, from startups to large enterprises.

Key Features of LLaMA

- **Generative AI Capabilities:** LLaMA excels at creating coherent, human-like text, enabling teams to draft reports, summaries, and updates with ease. Its generative capabilities are invaluable for streamlining communication and enhancing creativity.
- **Contextual Understanding:** By maintaining context over extended interactions, LLaMA ensures that responses and outputs remain accurate and relevant to the project at hand.
- **Multilingual Support:** LLaMA supports a wide range of languages, making it an ideal tool for teams operating in diverse linguistic environments.
- **Efficiency and Scalability:** Optimized for lightweight performance, LLaMA is accessible even on systems with limited computational resources, making it cost-effective and versatile.

Applications in Project Management

- **Automating Documentation:** LLaMA can generate meeting notes, project reports, and stakeholder updates, significantly reducing the administrative load on project managers.
- **Enhancing Brainstorming:** By providing creative suggestions and alternative strategies, LLaMA adds value to ideation sessions.
- **Supporting Communication:** With its multilingual capabilities, LLaMA ensures smooth communication among global teams and stakeholders.
- **Summarizing Complex Information:** It condenses lengthy documents into concise summaries, allowing teams to quickly grasp key points and make informed decisions.
- **Analyzing Feedback:** LLaMA can process feedback from stakeholders or team members to identify common themes and actionable insights.

Pricing

Meta offers flexible pricing plans for LLaMA access to accommodate diverse organizational needs:

- **Free Tier:** The free tier allows for basic access to LLaMA's features with usage limitations, making it suitable for smaller teams or for testing its capabilities.
- **Pay-As-You-Go:** LLaMA's pay-as-you-go model provides scalability for teams with fluctuating usage requirements. Pricing is based on token consumption, with rates starting as low as $0.07 per

1 million tokens for input and $0.30 per 1 million tokens for output.

- **Premium Subscription Plans:** Meta also offers advanced subscription plans for heavy users or enterprises requiring enhanced functionality. These plans include:

  - ➤ **LLaMA Pro:** Priced at $29.99 per month, this plan includes priority access to advanced features, faster processing times, and increased token limits.
  - ➤ **Enterprise Solutions:** Tailored for large organizations, enterprise packages include dedicated support, custom integrations, and enhanced scalability. Pricing for enterprise solutions is determined on a case-by-case basis.

BLOOM

Bloom, an open-source large language model developed by Hugging Face and its collaborators, is designed to democratize AI by providing robust natural language processing capabilities to a wide audience. Built on transparency and collaboration, Bloom is a powerful tool for teams looking to integrate AI into project management without relying on proprietary platforms.

Key Features of Bloom

Multilingual Proficiency: Bloom supports over 46 languages and 13 programming languages, making it an excellent choice for global teams and diverse use cases.

Open-Source Framework: As an open-source model, Bloom allows teams to customize and fine-tune its capabilities according to their unique project requirements.

Scalability: Bloom's flexible architecture makes it suitable for a range of applications, from small-scale tasks to large enterprise-level projects.

High-Quality Text Generation: Bloom produces coherent, human-like text outputs, making it ideal for drafting reports, summaries, and project updates.

Applications in Project Management

- **Automating Documentation:** Bloom can draft meeting notes, progress reports, and stakeholder communications, reducing administrative workload.
- **Enhancing Collaboration:** With its multilingual capabilities, Bloom enables seamless communication across global teams and stakeholders.
- **Summarizing Information:** It can condense lengthy project proposals or research documents into concise summaries for quick decision-making.
- **Idea Generation:** During brainstorming sessions, Bloom provides innovative ideas and alternative strategies, fostering creativity within the team.
- **Processing Feedback:** Bloom analyzes text-based feedback from team members or stakeholders, identifying key themes and actionable insights.

Pricing

Bloom offers flexible pricing options to cater to diverse user needs:

- **Free Tier:** The free tier allows users to send up to 50 messages, making it suitable for small teams or for testing the platform's capabilities.
- **Personal Account:** For $9.99 per month, this plan provides unlimited conversations, personalized experiences, and advanced features such as aligning to learning styles, synthesizing across chats, and building rich context.
- **Custom Group Accounts:** Tailored for larger teams or institutions, these accounts include features like classroom setups, study group tools, school licensing, and organizational deployments. Pricing for group accounts varies based on specific requirements and is determined on a case-by-case basis.

By offering a robust and accessible platform, Bloom empowers organizations to harness the power of AI in project management while maintaining control and flexibility over its use.

Foundational AI technologies like GPT, Gemini, LLaMA, and Bloom are revolutionizing project management by offering tools that optimize efficiency, enhance communication, and empower decision-making. These models have demonstrated their capabilities to transform how tasks are executed, allowing project managers to focus on strategy, creativity, and leadership rather than mundane, repetitive operations.

While each AI model has unique strengths – such as GPT's adaptability, Gemini's multimodal capabilities, LLaMA's scalability, and Bloom's open-source accessibility – they all share the common goal of streamlining workflows and fostering collaboration. Their integration into project management workflows has become a necessity for organizations aiming to stay competitive in an increasingly digital world.

The real power of AI lies in its ability to augment human expertise rather than replace it. By automating routine tasks, generating insights from data, and providing real-time assistance, these tools empower teams to tackle complex challenges with greater agility. However, their successful adoption requires a balanced understanding of both their potential and their limitations, as well as a commitment to ethical and responsible use.

As you move forward, keep in mind that AI is a tool that amplifies your abilities. Choosing the right AI technology for your team depends on your specific needs, resources, and goals. The subsequent chapters will guide you through practical steps to integrate these tools effectively into your workflows, ensuring that your projects achieve unprecedented levels of success and efficiency.

Chapter 3: AI-Enhanced Agile Methodologies

Agile methodologies have become the cornerstone of modern project management, providing teams with the flexibility to adapt quickly to changes, deliver value incrementally, and foster collaboration among stakeholders. Unlike traditional project management approaches, Agile prioritizes iterative development, continuous feedback, and a focus on delivering high-priority outcomes. Its principles have proven invaluable in industries ranging from software development to marketing, construction, and beyond.

However, despite its advantages, Agile frameworks are not without challenges. Manual backlog prioritization often relies on subjective judgment, which can lead to inefficiencies or missed opportunities. Resource limitations, such as uneven workloads or skill mismatches, can strain teams and compromise project outcomes. Additionally, Agile's demand for continuous adaptability can make it difficult to predict and mitigate risks effectively, especially in complex or fast-paced environments.

This is where artificial intelligence (AI) comes into play, offering a transformative opportunity to enhance Agile practices. AI can analyze vast amounts of data to provide actionable insights, automate repetitive tasks, and identify patterns that human teams might overlook. By integrating AI into Agile methodologies, project managers can optimize processes such as sprint planning, backlog management, and risk mitigation while empowering teams to focus on creativity and innovation.

In this chapter, we'll explore how AI can address these challenges and elevate Agile practices to new heights. From intelligent sprint planning to real-time project adaptation, you'll discover practical applications of AI that make Agile more efficient, effective, and adaptive to the demands of modern project management. By the end of this chapter, you'll have a clear understanding of how to seamlessly integrate AI into your Agile workflows, ensuring your projects stay on track and deliver exceptional results.

Adapting Traditional Agile Frameworks with AI

Agile methodologies thrive on responsiveness, flexibility, and iterative planning. These principles ensure teams can quickly adapt to changing requirements and consistently deliver value to stakeholders. However, the traditional implementation of Agile frameworks like Scrum and Kanban often relies heavily on human-driven processes. While effective, these processes can sometimes fall short in terms of speed, accuracy, and the ability to foresee risks in complex or data-intensive projects. This is where AI can step in as a powerful complement, enhancing Agile frameworks without compromising their core principles.

Identifying Areas Where AI Can Complement Agile Principles

AI aligns seamlessly with Agile's key tenets by automating repetitive tasks, analyzing vast datasets for actionable insights, and enhancing decision-making in real time. Key areas of synergy include:

- **Responsiveness:** AI-powered tools can monitor project metrics continuously, flagging delays or risks as they arise. This enables teams to make

quick adjustments without waiting for the next sprint review.
- **Flexibility:** AI tools offer predictive analytics that help anticipate changes in project scope, resource needs, or stakeholder priorities, making it easier to adapt plans dynamically.
- **Iterative Planning:** With AI's capacity to process historical and real-time data, teams can refine sprint goals and backlog priorities more effectively between iterations.

Examples of AI Integration with Popular Agile Frameworks

- **Scrum:**
  AI can enhance Scrum practices by optimizing sprint planning and automating tasks like assigning story points based on historical data. During sprint execution, AI tools monitor task progress, predict potential bottlenecks, and recommend course corrections in real time. For example, predictive analytics can help Scrum Masters identify risks before they escalate, improving team efficiency and project outcomes.
- **Kanban:**
  In Kanban systems, AI can analyze workflow patterns to predict task completion times and suggest adjustments to improve throughput. By monitoring work-in-progress (WIP) limits, AI ensures tasks flow smoothly across the board, reducing bottlenecks and enhancing team productivity. Visualizing these insights in dynamic Kanban boards helps teams focus on what matters most.

Benefits of Blending AI Capabilities with Agile

Integrating AI into Agile frameworks offers transformative benefits:

- **Data-Driven Decisions:** AI provides project managers with actionable insights derived from historical and real-time data, enabling more accurate prioritization, resource allocation, and risk mitigation.
- **Proactive Risk Management:** AI tools can identify potential risks early by analyzing patterns in project metrics, allowing teams to address issues before they disrupt progress.
- **Enhanced Productivity:** By automating time-consuming tasks such as backlog grooming or sprint reporting, AI frees up team members to focus on strategic and creative aspects of the project.
- **Improved Stakeholder Engagement:** AI-driven sentiment analysis can process stakeholder feedback to identify concerns or preferences, ensuring that project outputs align more closely with expectations.

Embracing AI within Agile frameworks allows project teams to enhance their ability to respond to challenges and opportunities, delivering faster, smarter, and more reliable results. The combination of human ingenuity and AI-powered insights ensures that Agile methodologies remain a robust and adaptable approach to modern project management.

AI-Powered Sprint Planning

Sprint planning is a crucial part of Agile methodologies, setting the stage for teams to align on priorities, define goals, and establish timelines for delivering value within a specific iteration. However, traditional sprint planning often involves subjective estimations, manual task prioritization, and the challenge of predicting potential risks. By incorporating AI into sprint planning, teams can leverage data-driven insights to create more efficient, accurate, and productive plans.

Leveraging AI to Analyze Historical Project Data

AI excels at processing historical project data, offering insights that improve sprint planning accuracy. By analyzing past sprints, AI tools can identify trends, such as recurring delays, high-performing team members, or tasks that consistently exceed time estimates. These insights enable project managers to set realistic goals and allocate resources effectively.

For example, AI might highlight that certain types of tasks historically take longer than expected, prompting teams to adjust their estimations. It can also identify which team members excel at specific types of work, allowing tasks to be assigned more strategically.

Automated Estimation of Task Durations and Dependencies

Accurate estimation is a critical aspect of sprint planning, but it's often challenging to predict how long tasks will take, especially for complex projects. AI can automate this process by analyzing task attributes, historical completion times, and dependencies to provide precise duration estimates.

For instance, an AI tool might recognize that a task involving specific integrations typically requires additional testing time. It could also map task dependencies automatically, ensuring that critical-path items are prioritized and potential blockers are addressed early.

Tools That Prioritize Tasks Based on Complexity, Team Capacity, and Deadlines

AI-powered tools can dynamically prioritize tasks by evaluating factors such as:

- **Task complexity:** Assigning simpler tasks alongside more challenging ones to balance workloads.
- **Team capacity:** Ensuring team members are not overburdened while maximizing productivity.
- **Deadlines:** Prioritizing time-sensitive tasks to meet stakeholder expectations.

These tools go beyond static priority lists, updating task rankings in real-time as team capacity or project scope changes. This ensures the sprint backlog remains optimized throughout the iteration.

AI's Role in Identifying Potential Bottlenecks Before They Occur

One of AI's most significant contributions to sprint planning is its ability to predict potential bottlenecks. By analyzing data like task dependencies, resource availability, and past performance, AI can flag areas where delays are likely to occur.

For example, AI might detect that a critical resource is overallocated, suggesting adjustments to prevent slowdowns. It could also simulate different sprint scenarios, helping teams identify the optimal sequence of tasks to minimize disruptions.

Benefits of AI-Powered Sprint Planning

- **Increased Accuracy:** Data-driven insights replace guesswork, improving task estimation and resource allocation.
- **Proactive Risk Management:** Early identification of bottlenecks reduces the likelihood of missed deadlines or scope changes.
- **Enhanced Team Productivity:** Optimized task prioritization ensures team members focus on high-value work.
- **Better Stakeholder Alignment:** Realistic sprint goals and timelines improve transparency and trust with stakeholders.

By leveraging AI-powered sprint planning tools, Agile teams can shift their focus from administrative tasks to delivering meaningful outcomes. These technologies not only enhance planning accuracy but also empower teams to adapt dynamically, ensuring every sprint contributes to the broader project vision.

Intelligent Backlog Management

Backlog management is at the heart of Agile workflows, serving as the foundation for sprint planning and prioritization. Traditionally, managing backlogs involves a mix of team discussions, stakeholder inputs, and manual organization. While effective to some extent, this process can become time-consuming and prone to oversight, especially for large or complex projects. By incorporating AI into backlog management, teams can dynamically prioritize tasks, refine items based on real-time feedback, and optimize workflows with greater efficiency and accuracy.

Using AI to Dynamically Prioritize User Stories and Tasks

AI-powered tools bring intelligence to backlog prioritization by evaluating user stories and tasks based on key factors such as business value, team capacity, and urgency. Unlike manual prioritization, which often relies on subjective judgment, AI uses objective data to rank backlog items.

For example, an AI tool can assess the potential impact of a user story by analyzing its alignment with business goals, expected ROI, and stakeholder requirements. It can also account for team availability and workload, ensuring that high-priority items are assigned to the right people without overburdening anyone. This dynamic prioritization ensures that the team always focuses on delivering the most value.

Sentiment Analysis for Stakeholder Feedback

Stakeholder feedback is a critical input for refining backlog items, but manually processing large volumes of feedback can be challenging. AI-powered sentiment analysis simplifies this task by analyzing feedback text to detect underlying emotions, concerns, and preferences.

For instance, if multiple stakeholders provide feedback on a feature, the AI can analyze the tone and keywords to identify whether the feature is meeting expectations or needs improvement. Based on this analysis, backlog items can be adjusted to reflect stakeholder priorities more effectively, ensuring alignment with project goals.

Automated Identification of Duplicate or Outdated Tasks

Over time, backlogs can become cluttered with duplicate or outdated tasks, creating confusion and inefficiency. AI tools excel at identifying and flagging such items automatically.

For example, an AI system might detect that two tasks in the backlog have similar descriptions or objectives, suggesting they could be merged. Similarly, it can flag tasks that haven't been updated in a long time, prompting the team to reevaluate their relevance. By decluttering the backlog, AI helps teams maintain focus on tasks that genuinely contribute to project success.

Visualizing and Reorganizing Backlogs Using AI-Driven Insights

AI-powered visualization tools transform the way teams interact with their backlogs. Instead of static lists, teams can view their backlogs as dynamic, interactive dashboards that highlight task priorities, dependencies, and progress.

For example, an AI tool might generate a heatmap showing which tasks are most critical based on deadlines and business value. It could also create dependency maps that reveal how changes to one item affect others in the backlog. This level of insight enables teams to reorganize their backlogs with confidence, ensuring that every item is positioned for maximum impact.

Benefits of AI-Driven Backlog Management

- **Improved Efficiency:** Automated prioritization and decluttering save time and reduce manual effort.

- **Enhanced Decision-Making:** Data-driven insights ensure that backlog adjustments are strategic and aligned with business goals.
- **Better Stakeholder Alignment:** Sentiment analysis integrates stakeholder feedback more effectively into the prioritization process.
- **Clarity and Focus:** AI-powered visualizations provide teams with a clear understanding of backlog priorities and dependencies.

Real-Time Project Adaptation Using AI Insights

One of the most significant advantages of integrating AI into Agile project management is the ability to adapt dynamically to evolving project conditions. Traditional Agile frameworks rely on regular meetings, such as daily standups or sprint reviews, to identify and address issues. While effective, these processes are inherently reactive, often identifying problems only after they've occurred. AI, on the other hand, enables real-time project adaptation by continuously monitoring progress, analyzing data, and providing actionable insights that help teams stay proactive.

Continuous Monitoring of Project Metrics

AI tools excel at tracking key project metrics, such as task completion rates, resource utilization, and timeline adherence in real time. By monitoring these indicators, AI can detect deviations from the plan as they happen, allowing teams to address issues immediately.

For example, an AI-powered project management tool like **Smartsheet with AI Insights** might notice a drop in task completion rates midway through a sprint. The tool could identify the underlying cause – such as an overburdened team member or a dependency delay – and suggest corrective actions, such as redistributing tasks or adjusting priorities.

Predictive Analytics for Risk Detection

AI's predictive analytics capabilities enable it to foresee potential bottlenecks, delays, or risks before they become critical issues. By analyzing historical data and current project conditions, AI can generate risk forecasts that help project managers make informed decisions.

For instance, an AI tool might flag that a high-priority task is at risk of delay due to insufficient resources. It could then recommend reassigning additional team members or extending the timeline for dependent tasks to avoid cascading delays.

- **Example:** In a website redesign project, the AI detects that a critical dependency – content delivery from a client – is running late. It suggests reallocating the team's focus to front-end development tasks that can be completed independently, minimizing downtime.

Dynamic Resource Allocation

As project conditions evolve, resource needs often shift. AI tools can adapt resource allocation in real time, ensuring optimal team performance and minimizing inefficiencies.

For example, if a team member unexpectedly takes leave, an AI-powered system can immediately assess the impact and suggest reassignments or workload adjustments. This ensures the sprint continues with minimal disruption.

- **Example Tool: ClickUp with AI Automations** can dynamically allocate tasks to available team members based on skill sets and capacity, ensuring that no task is left unattended.

Real-Time Feedback Loops

AI enhances feedback loops by continuously analyzing project data and providing actionable insights. These insights can inform daily standups, ensuring that teams address emerging issues proactively.

For instance, during a sprint, an AI tool might generate insights about a recurring issue with task handoffs between developers and testers. The project manager can bring this to the team's attention during the next standup, enabling them to address the issue immediately and improve workflow efficiency.

Example: Adapting to Scope Changes Mid-Sprint

Imagine a software development team working on a mobile app. Mid-sprint, a major stakeholder requests the addition of a new feature that wasn't part of the original plan. Traditionally, such a change could disrupt the sprint and affect deliverables.

With AI insights, the project manager can quickly assess the impact of the scope change. The AI tool evaluates:

- Which tasks can be deprioritized or deferred to the next sprint.
- How the new feature aligns with current sprint goals.
- Resource adjustments needed to accommodate the change.

The team uses the AI's recommendations to update the sprint backlog and reallocate resources, ensuring that the new feature is integrated without jeopardizing existing commitments.

Benefits of Real-Time Project Adaptation with AI

- **Proactive Problem-Solving:** Teams can address issues before they escalate, reducing delays and improving overall efficiency.
- **Increased Agility:** AI allows for quick and informed adjustments to plans, ensuring projects stay aligned with stakeholder priorities.
- **Optimized Resource Utilization:** Dynamic resource allocation ensures that team capacity is used effectively, even in the face of unexpected changes.
- **Enhanced Collaboration:** Real-time insights improve communication and decision-making across teams, fostering a culture of transparency and adaptability.

By leveraging AI for real-time project adaptation, Agile teams can navigate the complexities of modern projects with confidence, ensuring that deliverables remain on track while accommodating evolving requirements.

Prompt Engineering for Agile Team Collaboration

Prompt engineering, the art of crafting effective inputs for AI tools to generate accurate and actionable outputs, has emerged as a powerful skill in modern project management.

In the context of Agile, where collaboration and iterative feedback are essential, prompt engineering ensures that AI tools can contribute meaningfully to team discussions, decision-making, and task execution. By creating well-structured and contextually relevant prompts, Agile teams can unlock the full potential of AI, streamlining workflows and enhancing collaboration.

What is Prompt Engineering?

Prompt engineering involves designing specific and targeted queries or instructions for AI tools to generate desired outputs. In Agile teams, this could range from summarizing a sprint retrospective to brainstorming solutions for a roadblock. The goal is to communicate clearly with the AI, ensuring its responses align with the team's needs and goals.

Key Principles of Prompt Engineering for Agile Teams

1. Clarity and Specificity:
Prompts should clearly state the desired outcome or task. For example, instead of asking, "What are the blockers?" a better prompt might be, "Summarize the blockers identified in today's sprint retrospective and suggest possible solutions."

2. Contextual Relevance:
Providing context helps the AI generate accurate outputs. For instance, including details about the sprint's goals or the project timeline ensures that AI recommendations are tailored to the team's specific situation.

3. Iterative Refinement:
Prompts can be refined based on the AI's responses. Agile teams should treat prompt engineering as an iterative process, fine-tuning prompts until the desired output is achieved.

Applications of Prompt Engineering in Agile Collaboration

1. Brainstorming User Stories or Epics:
AI tools can assist teams in generating creative and comprehensive user stories by responding to well-structured prompts.

- Example Prompt: "Generate three user stories for a task management app targeting remote teams. Include acceptance criteria for each story."

2. Summarizing Sprint Retrospectives:
After sprint reviews, teams can use AI to distill key takeaways, enabling them to focus on actionable improvements.

- Example Prompt: "Summarize the key points from today's sprint retrospective, highlighting what went well, what didn't, and proposed actions for the next sprint."

3. Enhancing Team Communication:
AI can draft emails, meeting agendas, or status updates based on specific team needs.

- Example Prompt: "Create a weekly update email for stakeholders summarizing the sprint progress, completed tasks, and pending items."

4. Resolving Conflicts or Roadblocks:
AI tools can provide suggestions for addressing team challenges by responding to focused prompts.

- Example Prompt: "Suggest solutions for improving collaboration between the development and QA teams during testing phases."

5. Planning Sprint Goals:
AI can assist in defining and prioritizing sprint objectives by analyzing backlog items and stakeholder inputs.

- Example Prompt: "Recommend three sprint goals based on the current backlog items and the project's deadline in three months."

Tools for Prompt Engineering in Agile Teams

Several AI-powered tools are designed to respond effectively to prompts and facilitate Agile collaboration:

- **ChatGPT or GPT-based Assistants:** Ideal for brainstorming, summarizing, and drafting Agile artifacts.
- **Notion AI:** Helps Agile teams organize and refine ideas within shared workspaces.
- **Microsoft Copilot for Teams:** Enhances collaboration by generating insights and summaries during team discussions.

Benefits of Prompt Engineering for Agile Collaboration

- **Increased Efficiency:** Well-crafted prompts reduce the time spent rephrasing queries or refining AI outputs.
- **Improved Communication:** AI-generated summaries and updates ensure clarity and alignment across teams and stakeholders.

- **Enhanced Creativity:** Prompts encourage AI to contribute innovative ideas, improving brainstorming and problem-solving sessions.
- **Scalability:** Prompt engineering enables AI to support both small teams and larger, cross-functional groups effectively.

Best Practices for Prompt Engineering in Agile Teams

1. **Start Simple:** Begin with basic prompts and refine them based on the AI's response.
2. **Provide Context:** Include relevant project details to guide the AI's output.
3. **Be Iterative:** Treat prompt crafting as a continuous improvement process, similar to Agile principles.
4. **Test and Learn:** Experiment with different phrasing to identify what works best for your team's needs.

By mastering prompt engineering, Agile teams can harness AI as a valuable collaborator, driving greater efficiency, creativity, and alignment in every stage of the project lifecycle.

Solving Sprint Planning Challenges with an AI-Powered Tool

Imagine a software development team tasked with delivering a new feature-rich mobile app within a tight six-month deadline. The project is complex, involving multiple cross-functional teams and interdependent tasks. During their initial sprint planning session, the team encounters several challenges:

- **Inconsistent task estimation:** Team members struggle to agree on the time required for each task, leading to discrepancies in the sprint backlog.
- **Overloaded resources:** Several key team members are already at capacity, but no clear visibility exists on how to redistribute tasks effectively.
- **Identifying dependencies:** Manually mapping out task dependencies proves time-consuming and prone to errors, creating potential risks for the sprint timeline.

To address these issues, the project manager decides to use an AI-powered sprint planning tool, such as **Jira Align with AI Insights**[4] or **Monday.com's AI Planning Assistant**.[5] Here's how the tool transforms the process:

Step 1: Task Estimation and Prioritization

The team inputs a list of user stories and tasks into the AI-powered tool. The tool analyzes historical data from past sprints, including time spent on similar tasks and completion rates, to provide accurate time estimates for each item.

For instance, the AI determines that a task like "Develop login API" typically takes three days based on past sprint data, while a more complex task like "Implement payment gateway integration" requires eight days. The system ranks tasks by priority, factoring in business value, deadlines, and potential risks.

---

[4]https://www.atlassian.com/software/jira/align
[5]https://monday.com/w/ai

- **Benefit:** The team achieves consensus on task durations without lengthy debates, creating a realistic sprint backlog.

Step 2: Resource Optimization

Using predictive analytics, the AI evaluates the current workload of each team member, along with their skill sets and availability. The tool flags overloaded developers and suggests reallocating certain tasks to underutilized team members with matching skills.

For example, the AI recommends assigning a front-end task, originally allocated to an overburdened senior developer, to a junior developer who has demonstrated proficiency in similar tasks.

- **Benefit:** Workload distribution becomes balanced, ensuring no team member is overworked while maintaining efficiency.

Step 3: Mapping Dependencies

The AI identifies dependencies between tasks and visualizes them on a dynamic dependency map. For instance, it highlights that the "Database schema design" task must be completed before starting "Develop login API." It also flags potential bottlenecks, such as the delayed delivery of external APIs from a third-party vendor.

- **Benefit:** By visualizing dependencies, the team minimizes risks and plans tasks in the optimal sequence, avoiding delays later in the sprint.

Step 4: Real-Time Adjustments

Mid-sprint, the project scope changes when a major client requests a new feature be prioritized. The AI tool reassesses the sprint backlog in real time, adjusting task priorities and providing updated estimates. It also suggests tasks that can be deferred to the next sprint without impacting project deadlines.

For example, the AI shifts focus to implementing the new feature, deferring a lower-priority task like "Optimize image loading times" to the following sprint.

- **Benefit:** The team adapts to changing requirements seamlessly without jeopardizing sprint objectives or morale.

Outcome

By leveraging the AI-powered sprint planning tool, the team completes the sprint on schedule, with all critical tasks delivered. The project manager reports significant improvements in planning accuracy, resource utilization, and team satisfaction.

Tools Used in the Example

- **Jira Align with AI Insights:** Offers predictive analytics for task estimation, resource allocation, and risk identification.
- **Monday.com AI Planning Assistant:** Provides real-time adjustments to sprint backlogs and visualizes dependencies for better prioritization.

- **Trello AI Add-Ons:** Automates task tracking and sends notifications to prevent delays in dependent tasks. [6]

This example illustrates how AI-powered sprint planning tools transform challenges into opportunities for increased efficiency and collaboration. By adopting such tools, Agile teams can ensure that every sprint contributes meaningfully to the overall project vision.

---

[6]https://trello.com/power-ups/5fc56ddb4a80fe2c77acaf95/strategy-ai

Chapter 4: Strategic Planning and Vision Setting

Strategic planning and vision setting are the cornerstones of successful project management. Without a clear vision, even the most talented teams can find themselves navigating without direction, struggling to align their efforts with the organization's broader goals. This chapter explores how artificial intelligence can elevate strategic planning, providing tools to analyze complex data, anticipate risks, and create comprehensive project roadmaps that inspire and guide teams toward success.

The traditional approach to strategic planning often involves extensive manual effort – compiling data from multiple sources, conducting endless brainstorming sessions, and relying on subjective insights to forecast outcomes. While effective in some cases, this process can be slow, error-prone, and overly reliant on past experiences. AI changes the game by offering speed, precision, and the ability to uncover hidden patterns and trends that might otherwise go unnoticed.

AI-powered tools enable project managers to generate data-driven project visions, simulate multiple scenarios, and assess risks with remarkable accuracy. These technologies don't just enhance efficiency – they also empower teams to approach strategic planning with confidence, ensuring that every decision is grounded in actionable insights.

In this chapter, you'll discover how AI can assist in crafting a compelling vision that aligns with organizational objectives, inspires stakeholders, and motivates teams. From using AI to analyze competitive landscapes to creating detailed vision documents, we'll explore practical applications that bridge the gap between ambition and execution. You'll also gain access to prompt templates that enable you to extract maximum value from AI tools, ensuring that your strategic plans are not only innovative but also achievable.

Using AI for Strategic Analysis

Strategic analysis is the process of assessing an organization's internal and external environments to identify opportunities, anticipate risks, and chart a clear path toward achieving its goals. Traditionally, strategic analysis requires gathering extensive data, conducting in-depth research, and synthesizing insights manually – an often time-consuming and error-prone endeavor. Artificial intelligence (AI) revolutionizes this process by automating data collection, analyzing vast datasets, and generating actionable insights in real time.

How AI Transforms Strategic Analysis

AI excels at processing large volumes of data from diverse sources, such as market trends, competitor activities, and organizational performance metrics. By identifying patterns and correlations that might be overlooked by human analysis, AI enhances the quality and speed of strategic decision-making.

Automating Data Collection and Analysis
One of AI's greatest strengths is its ability to gather and process data at a scale and speed unmatched by traditional methods. AI-driven platforms can continuously monitor sources such as industry reports, social media discussions, economic indicators, and financial performance data. This eliminates the need for manual data aggregation and ensures

that strategic decisions are based on the most up-to-date and comprehensive information available.

For example, natural language processing (NLP) tools can scan thousands of market reports and extract relevant insights in seconds. AI-powered business intelligence software, like Tableau and Power BI, can integrate real-time data feeds and generate interactive dashboards that provide executives with a clear picture of emerging trends.

Identifying Market Trends and Competitive Insights

AI algorithms can detect patterns in industry data that might take human analysts weeks or months to uncover. Machine learning models analyze consumer behavior, pricing fluctuations, and shifts in demand to predict market movements with a high degree of accuracy. This enables organizations to adapt their strategies proactively rather than reactively.

Competitive intelligence tools like Crayon and SEMrush leverage AI to track competitors' online presence, product launches, and advertising strategies. By analyzing these data points, project managers can make informed strategic choices that position their organization ahead of the competition.

Enhancing SWOT Analysis with AI

The traditional SWOT (Strengths, Weaknesses, Opportunities, and Threats) analysis is often subjective, relying on human judgment and limited data inputs. AI enhances SWOT analysis[7] by providing data-driven insights that minimize biases and uncover hidden opportunities and threats.

---

[7] https://prometai.app/blog/ai-swot-analysis-evolution

For instance, AI-powered analytics can identify internal performance trends by analyzing employee productivity, project success rates, and resource utilization. Externally, AI can assess risks such as economic downturns, geopolitical factors, or shifts in consumer preferences by processing global data sources in real-time. This level of analysis ensures that strategic planning is grounded in objective, actionable intelligence.

Predictive Analytics for Risk and Opportunity Assessment
AI-driven predictive analytics leverage historical data to forecast future scenarios, helping organizations mitigate risks and seize opportunities before they arise. These models use statistical techniques and machine learning to generate probability-based insights into potential market disruptions, supply chain vulnerabilities, and customer behavior shifts.

For example, AI can simulate the impact of various strategic decisions, such as entering a new market or launching a new product. These simulations provide leaders with multiple outcome scenarios, allowing them to choose the best course of action with confidence.

Sentiment Analysis for Stakeholder and Market Perception
AI-powered sentiment analysis tools assess public and stakeholder opinions by analyzing social media posts, news articles, and customer reviews. This helps organizations understand how their brand is perceived and identify areas for improvement in their strategic initiatives.

For example, a company planning a product launch can use AI to analyze online discussions and gauge customer expectations. If sentiment analysis reveals negative perceptions about a specific product feature, the organization can address concerns proactively before the launch, improving customer satisfaction and engagement.

The Strategic Advantage of AI-Driven Analysis
AI not only accelerates the strategic analysis process but also enhances the depth and reliability of insights. By automating data collection, identifying emerging trends, and predicting future scenarios, AI empowers organizations to make more informed and proactive strategic decisions. As AI technology continues to evolve, project managers who leverage its capabilities will gain a significant competitive advantage in setting clear, data-driven visions for their organizations.

Generating Comprehensive Project Vision Documents

A well-crafted project vision document is essential for aligning teams, setting clear objectives, and ensuring that all stakeholders share a common understanding of the project's purpose and direction. Traditionally, creating these documents requires extensive brainstorming, stakeholder consultations, and iterative revisions. AI significantly streamlines this process by providing structured insights, automating content generation, and ensuring that project vision documents remain data-driven and actionable.

The Importance of a Project Vision Document
A project vision document serves as the foundation for successful execution. It typically includes:

- **Project Purpose** – The reason behind initiating the project and the problem it aims to solve.

- **Goals and Objectives** – Clearly defined success criteria and measurable outcomes.

- **Stakeholder Alignment** – A shared understanding among project teams, executives, and external stakeholders.

- **Scope and Constraints** – Boundaries of the project to prevent scope creep.

- **Strategic Fit** – How the project aligns with broader organizational goals and market needs.

With AI, creating a vision document becomes a structured, intelligent process rather than a purely manual effort.

How AI Enhances Vision Document Creation
Let's take a more in depth look at the ways in which AI can help project managers enhance vision documents and improve upon them.

Structuring and Automating Document Drafting
AI-powered writing assistants, such as GPT-based tools, can generate detailed project vision documents from simple prompts. By feeding an AI tool with project details – such as key objectives, stakeholders, and industry trends – it can draft a comprehensive document in minutes.

For example, a prompt like:

"Generate a project vision document for a software development initiative aimed at automating customer service through AI-powered chatbots. Include objectives, expected benefits, and alignment with business goals." can produce a well-organized draft, saving significant time and effort.

AI-driven platforms like Notion AI, Jasper, and ChatGPT can further refine sections based on real-time feedback, ensuring clarity and completeness.

Data-Driven Goal Definition

AI excels in setting realistic and data-backed project goals. Instead of relying solely on intuition, AI can:

- Analyze past project performance to recommend achievable milestones.

- Compare industry benchmarks to suggest competitive yet feasible objectives.

- Assess risk factors and adjust goals accordingly.

For example, an AI-powered business intelligence tool might identify that similar projects in the industry achieve a 20% cost reduction through automation. The project vision document can then incorporate this benchmark as a key target.

Ensuring Stakeholder Alignment

One of the biggest challenges in project planning is aligning various stakeholders who may have different expectations. AI can:

- Analyze stakeholder input from emails, surveys, and meetings to identify common themes.

- Generate stakeholder-specific summaries highlighting relevant benefits and concerns.

- Use sentiment analysis to detect areas of disagreement early and suggest strategies for consensus-building.

For instance, AI can summarize executive leadership priorities alongside frontline team concerns and integrate them into the vision document, ensuring everyone's needs are addressed.

Automating Visuals and Infographics

A project vision document is more effective when it includes compelling visuals. AI-powered tools can generate:

- **Mind maps** that visually outline project objectives and dependencies.

- **Infographics** summarizing key project benefits and strategic fit.

- **Timelines** predicting project phases and major milestones based on historical data.

Tools like Canva AI and Lucidchart AI automate the creation of these elements, making vision documents more engaging and easier to understand.

Scenario Planning and Risk Assessment Integration

AI-powered simulations can help project managers craft vision documents that account for multiple scenarios. Instead of presenting a rigid plan, AI can:

- Generate best-case, worst-case, and most likely projections for project outcomes.

- Predict potential risks based on historical data and industry trends.

- Suggest risk mitigation strategies that can be embedded directly into the document.

For example, if AI identifies a 30% chance of vendor delays due to supply chain issues, the project vision document can include contingency plans for alternative sourcing strategies.

The Advantages of Using AI for Vision Documents

AI transforms project vision document creation from a manual, time-consuming process into an automated, data-driven effort. By structuring content, setting realistic goals, aligning stakeholders, incorporating visuals, and predicting risks, AI ensures that vision documents serve as a strong foundation for project success.

With a solid vision in place, project teams can move forward with confidence, knowing that their objectives are clear, their risks are assessed, and their strategies are backed by intelligent insights.

Scenario Planning and Risk Assessment

In project management, uncertainty is inevitable. No matter how well-planned a project is, unforeseen obstacles can arise – changing market conditions, technological disruptions, regulatory shifts, or even internal resource constraints. Scenario planning and risk assessment help project managers prepare for these uncertainties by considering multiple possible futures and proactively developing strategies to navigate potential challenges.

Artificial intelligence (AI) enhances scenario planning and risk assessment by leveraging vast amounts of historical data, identifying emerging patterns, and generating predictive models. AI-powered tools can assess potential risks in real-time, simulate various future scenarios, and recommend optimal strategies for mitigating uncertainties. This section explores how AI can revolutionize risk management, ensuring that projects remain resilient and adaptable in a fast paced and changing environment.

The Role of AI in Scenario Planning

Scenario planning involves creating multiple "what-if" situations to anticipate how different internal and external factors might impact a project. AI strengthens this process by automating data analysis, simulating future outcomes, and recommending strategic responses.

Automating Historical Data Analysis

Traditionally, scenario planning requires extensive manual research – analyzing past projects, industry trends, and external factors to make informed predictions. AI simplifies this process by rapidly processing vast datasets, identifying patterns, and drawing actionable insights.

For example, AI can analyze previous project failures and successes to determine which factors contributed to delays or cost overruns. This information can then be used to create realistic scenario models that help project managers prepare for similar challenges in future projects.

AI-driven analytics platforms like Tableau, Power BI, and IBM Watson can compile and visualize historical data, allowing project managers to explore key trends and risks more effectively.

Generating Realistic Future Scenarios

AI-powered predictive models can simulate multiple future scenarios based on current data trends. These models consider factors such as:

- Market fluctuations
- Economic downturns or booms
- Changes in consumer behavior
- Regulatory or policy shifts
- Technological advancements
- Supply chain disruptions

By running simulations, AI enables project teams to evaluate best-case, worst-case, and most likely scenarios. This prepares them for different possibilities and ensures they have contingency plans in place.

For instance, a construction project using AI scenario modeling might anticipate material shortages due to global supply chain disruptions. AI can then generate alternate supply strategies or suggest budget adjustments to accommodate potential price increases.

Identifying Early Warning Signs

AI systems continuously monitor external signals – such as news articles, financial reports, and social media discussions – to detect early warning signs of potential disruptions. Natural language processing (NLP) tools scan real-time data sources and flag emerging risks, allowing project managers to react before an issue escalates.

For example, if AI detects increased discussions about cybersecurity threats within a certain industry, it can alert IT project teams to strengthen security measures before an attack occurs. This proactive approach ensures organizations stay ahead of potential challenges.

AI-Driven Risk Assessment

Risk assessment is a critical component of project management, helping teams identify, evaluate, and prioritize risks that could hinder project success. AI enhances risk assessment by providing real-time risk monitoring, intelligent risk scoring, and automated mitigation recommendations.

Real-Time Risk Monitoring and Detection

Unlike traditional risk assessments, which are often conducted periodically, AI enables continuous risk monitoring.[8] AI-powered tools scan vast amounts of project data, financial records, and industry reports to identify emerging risks as they develop.

For example, an AI-driven project management system like Smartsheet AI can detect delays in task completion and automatically flag potential project overruns. If a project milestone is at risk due to resource constraints, the AI can alert managers immediately, enabling them to reallocate resources before a delay becomes critical.

AI also plays a key role in compliance risk management. Platforms such as IBM OpenPages and SAS Risk Management use AI to monitor regulatory changes and ensure organizations remain compliant with evolving policies.

Intelligent Risk Scoring and Prioritization
Not all risks carry the same level of impact. AI-powered risk assessment tools assign risk scores based on factors such as probability, severity, and potential cost. These risk scores help project managers prioritize high-impact risks that require immediate attention.

For instance, AI models can assign risk scores to vendor partnerships based on past performance, financial stability, and geopolitical factors. A project team working with an overseas supplier might receive a high-risk score due to unstable political conditions in that region, prompting managers to consider alternative suppliers.

---

[8]https://medium.com/walmartglobaltech/ai-driven-continuous-monitoring-the-future-of-third-party-risk-management-01e40e789d99

Tools like RiskWatch and RiskLens leverage AI-driven risk quantification, allowing organizations to measure financial and operational risks with greater accuracy.

Automated Risk Mitigation Strategies

Once AI identifies risks, it can suggest data-driven mitigation strategies to reduce their impact. These recommendations may include:

- Adjusting project timelines to account for anticipated delays

- Allocating additional resources to high-risk tasks

- Diversifying supplier networks to reduce dependency on a single vendor

- Enhancing cybersecurity measures to prevent data breaches

- Implementing redundancy plans for critical infrastructure

For example, if AI predicts a high likelihood of employee burnout due to workload imbalances, it may suggest redistributing tasks among team members or adjusting deadlines to ensure sustainable productivity.

AI tools like Project Insight and Jira AI help automate these risk mitigation strategies, offering dynamic recommendations tailored to project-specific challenges.

AI in Crisis Response Planning

While risk assessment focuses on prevention, crisis response planning prepares organizations to react effectively when unexpected events occur. AI enhances crisis management by providing real-time decision support, automated response workflows, and post-crisis analytics.

AI-Driven Decision Support

During a crisis, AI-powered dashboards offer real-time insights to help teams make informed decisions under pressure. AI models can simulate the impact of different response strategies and recommend the most effective course of action.

For instance, during a cybersecurity breach, AI can quickly analyze system logs, identify the source of the attack, and suggest containment measures – such as isolating affected servers or implementing firewall updates.

Automated Emergency Workflows

AI can automate emergency response workflows, ensuring teams follow predefined protocols during crises. These workflows may include:

- Notifying key stakeholders and employees

- Activating backup systems or alternative supply chains

- Implementing data recovery measures

- Conducting impact assessments

Tools like ServiceNow AI and Splunk's AI-driven incident response ensure organizations react swiftly and efficiently to unexpected disruptions.

Post-Crisis Analysis and Learning

Once a crisis is resolved, AI tools analyze data from the event to extract key lessons and improve future preparedness. AI-driven reports highlight what went wrong, what worked well, and how response strategies can be refined.

For example, after a product recall, AI can analyze customer feedback, identify reputational damage, and suggest corrective actions to rebuild trust.

Key Benefits of AI in Scenario Planning and Risk Assessment

AI transforms scenario planning and risk assessment by providing real-time insights, predictive analytics, and automated decision support. By leveraging AI-powered tools, organizations can anticipate potential risks, develop contingency plans, and respond to crises more effectively.

Key benefits of AI in scenario planning and risk assessment include:

- Faster and more accurate risk identification
- Proactive problem-solving through predictive modeling
- Dynamic risk scoring for prioritization
- Automated mitigation and response strategies
- Continuous monitoring for emerging threats

By integrating AI into strategic planning, project managers gain unparalleled foresight and resilience, ensuring that projects remain adaptable and successful in an unpredictable world.

AI-Driven Competitive Intelligence Gathering

Staying ahead of the competition requires more than intuition and experience – it demands data-driven insights. Competitive intelligence (CI) is the process of gathering, analyzing, and applying information about competitors, industry trends, and market conditions to inform strategic decisions.[9] Traditionally, this process required labor-intensive research, manual data collection, and extensive industry networking.

Artificial intelligence (AI) is revolutionizing competitive intelligence by automating data collection, uncovering hidden patterns, and delivering actionable insights with unprecedented speed and accuracy. AI-powered tools enable businesses to monitor competitor activities, predict market shifts, and gain a strategic advantage without the exhaustive effort of traditional CI methods.

In this section, we will explore how AI enhances competitive intelligence, the tools available, and best practices for leveraging AI-driven insights effectively.

How AI Enhances Competitive Intelligence

AI-driven competitive intelligence operates on several levels, from tracking real-time competitor movements to analyzing vast amounts of industry data to uncover strategic opportunities. Here are some of the key ways AI enhances CI efforts:

Automated Data Collection and Processing

Gathering competitive intelligence manually is time-consuming and prone to human error. AI automates this process by scanning thousands of sources, including:

- Competitor websites

- Press releases

- Social media activity

- Online reviews and customer feedback

---

[9] https://www.investopedia.com/terms/c/competitive-intelligence.asp

- Financial reports

- Industry news and whitepapers

Natural language processing (NLP) algorithms allow AI to extract relevant insights from unstructured data sources, making sense of large volumes of text-based information. AI-powered web scraping tools like Crayon and SimilarWeb continuously monitor competitor websites, tracking changes in pricing, product offerings, and marketing strategies.

Real-Time Competitive Monitoring

AI-powered tools can provide real-time alerts when competitors make significant moves, such as launching a new product, changing their pricing model, or shifting their branding strategy. These alerts ensure businesses can respond quickly and strategically rather than being caught off guard.

For example, AI-driven social media analytics tools like Brandwatch or Sprinklr monitor competitors' mentions across platforms such as Twitter, LinkedIn, and Instagram. If a competitor is generating buzz due to a successful campaign, AI can analyze customer sentiment and help organizations craft counterstrategies to capitalize on emerging trends.

Predictive Market Trend Analysis

AI doesn't just track historical data – it predicts future trends by analyzing vast amounts of structured and unstructured data. Machine learning algorithms assess patterns in consumer behavior, economic indicators, and technological advancements to forecast industry shifts.

For instance, AI-powered platforms like AlphaSense use deep learning to analyze earnings calls, industry reports, and financial disclosures, identifying signals that indicate potential market disruptions. Businesses can then adjust their strategies to proactively position themselves ahead of industry shifts.

AI Tools for Competitive Intelligence Gathering

A variety of AI-powered platforms specialize in competitive intelligence, offering different functionalities tailored to business needs. Here are some of the most effective tools available today:

1. *Crayon*
   - **Function:** Competitive intelligence platform that tracks changes on competitor websites, pricing, and customer feedback.

   - **AI Capabilities:** Automates competitive monitoring and provides real-time alerts on industry shifts.

   - **Use Case:** A retail company using Crayon can receive notifications whenever a competitor updates their product offerings or adjusts pricing, allowing them to respond strategically.

2. *SimilarWeb*
   - **Function:** Provides website traffic analysis, digital market research, and audience insights.

   - **AI Capabilities:** Uses machine learning to analyze online behavior and predict trends in website engagement.

   - **Use Case:** A SaaS company can use SimilarWeb to compare their website traffic to competitors and identify opportunities to improve SEO and advertising strategies.

3. *Brandwatch*

- **Function:** Social media listening and consumer intelligence tool.

- **AI Capabilities:** Uses NLP and sentiment analysis to assess brand perception and track competitor mentions.

- **Use Case:** A marketing team can use Brandwatch to evaluate how customers react to a competitor's product launch and adjust their messaging accordingly.

4. *AlphaSense*

- **Function:** AI-powered market intelligence platform for financial and business research.

- **AI Capabilities:** Uses deep learning to analyze earnings reports, investor calls, and company filings.

- **Use Case:** A corporate strategy team can leverage AlphaSense to identify early warning signs of industry disruptions and adjust their investment decisions accordingly.

Key AI Techniques for Competitive Intelligence
Sentiment Analysis for Competitor Brand Perception

AI-driven sentiment analysis evaluates how consumers perceive competitors by analyzing online reviews, social media comments, and customer feedback. By understanding sentiment trends, businesses can identify areas where competitors are excelling – or struggling – and adjust their own strategies accordingly.

For example, if AI detects that a competitor's new product is receiving negative feedback about pricing, a company can emphasize affordability in its own marketing campaigns to attract dissatisfied customers.

AI-Powered SWOT Analysis
AI enhances SWOT (Strengths, Weaknesses, Opportunities, Threats) analysis by providing objective, data-driven insights. Instead of relying on subjective judgment, AI can:

- **Identify strengths** by analyzing positive customer reviews and product performance.

- **Uncover weaknesses** by tracking customer complaints and service issues.

- **Spot opportunities** by detecting emerging trends in consumer behavior.

- **Flag threats** by identifying competitor growth strategies and new market entrants.

AI tools like SEMrush integrate market analytics with competitor research to enhance strategic planning based on real-world data.

AI-Driven Competitive Pricing Analysis
AI-powered pricing tools analyze competitor pricing strategies in real-time and suggest optimal pricing models. Machine learning algorithms evaluate factors such as:

- Seasonal demand fluctuations

- Competitor price adjustments

- Consumer purchasing behavior

- Inventory levels

For example, an e-commerce business using an AI-powered pricing tool like Prisync can automatically adjust prices to stay competitive without sacrificing profit margins.

Best Practices for Using AI-Driven Competitive Intelligence

To maximize the effectiveness of AI-driven competitive intelligence, organizations should follow these best practices:

Combine AI with Human Expertise

AI can process vast amounts of data, but human analysts are needed to interpret findings, provide context, and make strategic decisions. The best approach combines AI's computational power with human intuition and industry knowledge.

Focus on Actionable Insights

Collecting competitive data is useless unless it leads to strategic action. Businesses should ensure that AI-generated insights directly inform marketing strategies, product development, and business planning.

Ensure Ethical Use of AI for Competitive Intelligence

AI-driven intelligence gathering must adhere to ethical and legal standards. Organizations should avoid unethical data collection practices, such as scraping confidential information or violating privacy regulations like GDPR.

Continuously Update AI Models

The competitive landscape is constantly evolving. AI models should be regularly updated with new data to ensure predictions remain accurate and relevant. Businesses should also refine their AI strategies based on performance feedback.

How AI is Transforming Competitive Intelligence

AI is transforming competitive intelligence by automating data collection, monitoring competitors in real time, predicting market trends, and uncovering hidden insights. By leveraging AI-powered tools such as Crayon, SimilarWeb, Brandwatch, and AlphaSense, businesses can stay ahead of the competition with minimal manual effort.

The integration of AI-driven sentiment analysis, SWOT analysis, and competitive pricing insights allows organizations to make informed, strategic decisions that ensure long-term success.

As AI continues to evolve, companies that embrace AI-powered competitive intelligence will gain a significant advantage in an increasingly dynamic and competitive market.

Prompt Templates for Strategic Planning

AI-powered tools can significantly enhance strategic planning by providing data-driven insights, automating complex analyses, and generating actionable recommendations. However, to maximize AI's potential, project managers must learn how to craft effective prompts that extract relevant and meaningful responses.

This section provides AI prompt templates for various strategic planning scenarios discussed in this chapter. These prompts can be used with AI-powered tools such as ChatGPT, Jasper, and Notion AI or business intelligence platforms like Tableau and AlphaSense. Each prompt is designed to help project managers streamline strategic analysis, scenario planning, risk assessment, and competitive intelligence gathering.

Prompts for Strategic Analysis
1. Identifying Key Market Trends

"Analyze the latest industry trends in [industry name] over the past [time period]. Identify emerging technologies, consumer behavior shifts, and economic factors that could impact strategic planning for a company operating in this industry."

2. SWOT Analysis for Project Planning

"Perform a SWOT analysis for a [industry name] company planning to launch a new product in [market name]. Identify key strengths, weaknesses, opportunities, and threats based on current industry data and market conditions."

3. Data-Driven Goal Setting

"Based on past performance data of similar projects, recommend three realistic and measurable goals for a [type of project]. Include benchmarks from competitors and industry best practices."

Prompts for Generating a Project Vision Document
4. Drafting a Project Vision Statement

"Create a compelling vision statement for a [project type] that aligns with the strategic goals of [company name]. Ensure the vision emphasizes innovation, stakeholder value, and long-term growth."

5. Aligning a Project with Organizational Strategy

"Analyze the strategic objectives of [company name] and suggest how the [specific project] can align with these goals. Provide three key ways the project contributes to the company's competitive advantage."

6. Creating a Roadmap for a New Initiative

"Generate a high-level roadmap for a [type of project] over a [time period]. Outline the major milestones, potential risks, and key performance indicators (KPIs) to measure success."

Prompts for Scenario Planning and Risk Assessment
7. Best-Case, Worst-Case, and Most-Likely Scenarios

"For a [project type] in [industry], generate three scenarios: best-case, worst-case, and most-likely outcomes. Include potential risks, financial impact, and mitigation strategies for each scenario."

8. Identifying Potential Project Risks

"Analyze the top five risks associated with a [project type] in [industry]. Rank them based on likelihood and impact, and suggest mitigation strategies."

9. AI-Powered Risk Mitigation Strategies

"Given a project with [specific risk factor], recommend three AI-driven mitigation strategies. Explain how predictive analytics and automation can help prevent or minimize this risk."

10. Predicting Market Disruptions

"Using historical data and current market trends, predict potential disruptions in [industry name] over the next [time period]. Provide insights on how a business can adapt to remain competitive."

Prompts for Competitive Intelligence Gathering
11. Monitoring Competitor Activity

"Track the latest strategic moves of [competitor name] in [industry name]. Summarize recent product launches, marketing campaigns, pricing changes, and customer sentiment over the past [time period]."

12. Benchmarking Against Industry Leaders

"Compare [company name] with the top five competitors in [industry name]. Identify key differentiators, strengths, and areas for improvement based on recent financial reports, market share, and innovation efforts."

13. Competitor Pricing Analysis

"Analyze the pricing strategy of [competitor name] compared to industry averages. Provide insights into how pricing adjustments could impact market positioning and customer acquisition."

14. Understanding Customer Sentiment Toward Competitors

"Perform a sentiment analysis of online reviews and social media discussions related to [competitor name]. Identify common themes in customer satisfaction, complaints, and brand perception."

15. Evaluating a Competitor's Strategic Direction

"Analyze the recent acquisitions, partnerships, and investments made by [competitor name] over the last [time period]. Provide insights on their likely strategic direction and potential impact on the industry."

Best Practices for Using AI-Generated Prompts Effectively

Be Specific

When crafting prompts, include as much detail as possible – such as industry name, competitor names, project types, and time frames. This ensures that AI provides highly relevant insights rather than generic responses.

Iterate and Refine

AI-generated outputs improve with refinement. If the initial response lacks depth or specificity, modify the prompt by adding more context or asking follow-up questions.

Combine AI Insights with Human Expertise

AI tools are powerful for generating data-driven insights, but human judgment is still essential. Always review AI-generated content and validate key findings before making strategic decisions.

Use AI to Explore Multiple Perspectives

Rather than relying on a single response, ask AI to generate different viewpoints. For example, requesting both an optimistic and a cautious forecast allows for more balanced decision-making.

Keep Ethical Considerations in Mind

When using AI for competitive intelligence, ensure that data collection follows legal and ethical guidelines. Avoid prompts that may encourage the misuse of proprietary or confidential information.

AI-powered strategic planning is most effective when project managers learn how to ask the right questions. The 15 prompt templates provided in this section cover essential aspects of project visioning, scenario planning, risk assessment, and competitive intelligence gathering. By using these prompts with AI tools, organizations can improve decision-making, anticipate risks, and gain a competitive edge in their industries.

AI is a powerful ally in strategic planning, but it should be used as a collaborative tool alongside human expertise. By mastering prompt engineering, project managers can unlock AI's full potential, ensuring that every strategic initiative is well-informed, proactive, and adaptable.

Chapter 5: Project Scoping and Requirements Gathering

Defining the scope and requirements of a project is one of the most critical stages in project management. A poorly scoped project can lead to scope creep, misaligned expectations, budget overruns, and missed deadlines. To ensure success, project managers must establish clear boundaries, define deliverables, and gather precise requirements from stakeholders. Traditionally, this process has been time-consuming, prone to misinterpretation, and highly dependent on manual input – but artificial intelligence (AI) is transforming how project scoping and requirements gathering are conducted.

AI-powered tools can analyze vast amounts of data, identify inconsistencies, and automate tedious documentation, allowing project managers to create well-defined project scopes with greater accuracy and efficiency. By leveraging AI for stakeholder analysis, requirement validation, and risk identification, organizations can minimize ambiguity, streamline communication, and ensure alignment from the outset.

In this chapter, we will explore how AI enhances each aspect of project scoping and requirements gathering, including:

- **AI-assisted requirements definition:** Automating the process of gathering, analyzing, and refining project requirements.

- **Intelligent stakeholder analysis:** Using AI to identify key stakeholders, map their influence, and understand their priorities.

- **Creating detailed project charters:** Leveraging AI to generate structured project charters that align with organizational goals.

- **Identifying potential project risks and mitigation strategies:** Utilizing AI-driven predictive analytics to assess risks before they escalate.

- **Prompt frameworks for comprehensive requirements documentation:** Providing structured AI prompts to ensure clarity, completeness, and consistency in project documentation.

By the end of this chapter, you will have a practical, AI-driven approach to defining project scope and gathering requirements with precision – laying the base work for successful project execution.

AI-Assisted Requirements Definition

Gathering and defining project requirements is a crucial step in ensuring a project's success. Requirements provide the foundation for planning, execution, and final deliverables, helping teams align expectations with stakeholders and avoid costly misunderstandings. However, traditional requirements gathering can be time-consuming, inconsistent, and prone to human bias or misinterpretation.

AI-driven tools are transforming how project managers define requirements by automating data collection, analyzing historical project data, and generating structured documentation with enhanced accuracy. AI not only accelerates the process but also minimizes errors, improves consistency, and helps identify gaps early in project planning.

How AI Enhances Requirements Definition

AI streamlines the process of gathering and defining project requirements through the following key capabilities:

Automating Requirement Collection from Multiple Sources

One of the biggest challenges in requirements gathering is compiling information from multiple stakeholders, documents, and data sources. AI-powered tools can aggregate requirements from:

- Past project documentation

- Industry standards and best practices

- Customer feedback and support tickets

- Meeting transcripts and email communications

- Competitor product specifications

For example, AI-driven tools like IBM Engineering Requirements Management DOORS and Jira AI can scan historical project data and automatically suggest initial requirement drafts. This reduces the manual effort required and ensures requirements are comprehensive and aligned with industry benchmarks.

NLP-Powered Stakeholder Interviews and Meeting Summaries

Traditionally, requirement gathering involves lengthy meetings and stakeholder interviews, which must then be manually documented and analyzed. AI-powered Natural Language Processing (NLP) tools can:

- Transcribe and summarize meetings in real time

- Extract key themes, action items, and stakeholder concerns

- Identify contradictions or inconsistencies in stakeholder inputs

- Suggest missing requirements based on common project patterns

For instance, tools like Otter.ai or Fireflies.ai can record stakeholder discussions, extract important takeaways, and generate structured requirement documents automatically. This eliminates misinterpretations and ensures that no crucial details are overlooked.

AI-Generated Requirement Drafting and Validation

AI-powered writing assistants can generate structured requirement documents based on project specifications, business needs, and historical data. These tools provide:

- **Predefined templates** tailored to different project types

- **Auto-generated requirement descriptions** based on minimal input

- **Real-time validation checks** to ensure clarity and feasibility

For example, AI tools like ChatGPT, Jasper AI, and Notion AI can take a simple prompt like: *"Generate functional and non-functional requirements for a mobile banking app, ensuring compliance with financial regulations and cybersecurity best practices."* and produce a detailed requirement document in minutes.

Identifying Gaps and Conflicts in Requirements

AI-powered requirement validation tools can compare project requirements against the following:

- Past successful projects

- Industry standards

- Legal and compliance guidelines

- Technical feasibility constraints

These tools flag inconsistencies, missing requirements, or potential conflicts before they become costly issues. For example, an AI tool analyzing software development requirements might detect that security measures are not defined or that a requirement contradicts existing system capabilities.

Platforms like ReqSuite RM and IBM Watson AI provide automated checks to ensure completeness, accuracy, and compliance before requirements are finalized.

Predictive Requirement Prioritization

AI can prioritize requirements based on impact, feasibility, and stakeholder needs. By analyzing past project data and market trends, AI models can:

- Rank requirements by business value and urgency

- Highlight dependencies between different requirements

- Suggest optimizations based on cost-benefit analysis

For example, an AI tool assessing software project requirements might prioritize core security features over UI enhancements based on risk factors and regulatory compliance. This helps project managers focus on high-impact deliverables first.

AI Tools for Requirements Definition

Here are some leading AI-powered tools that assist in defining and managing project requirements:

IBM Engineering Requirements Management DOORS

- Automates requirement collection and validation

- Detects conflicts and inconsistencies in requirements

- Ensures compliance with industry standards

Jira AI-Powered Requirement Management

- Uses AI to suggest missing or conflicting requirements

- Prioritizes tasks based on historical project data

- Integrates with Agile development workflows

Notion AI and ChatGPT

- Auto-generates structured requirement documents

- Refines stakeholder inputs into clear, concise specifications

- Provides real-time grammar and clarity checks

ReqSuite RM

- Uses AI to analyze and optimize requirement completeness

- Predicts potential project risks based on requirement gaps

- Offers smart requirement templates for different industries

AI-assisted requirements definition is transforming how project managers gather, validate, and document project needs. By automating data collection, summarizing stakeholder inputs, identifying gaps, and predicting requirement prioritization, AI significantly reduces manual effort, minimizes errors, and enhances the accuracy of project planning.

As AI tools continue to evolve, organizations that integrate AI-driven requirements management will benefit from faster project initiation, improved collaboration, and greater project success rates.

Intelligent Stakeholder Analysis

Stakeholder analysis is a crucial part of project scoping, as it helps project managers identify key individuals, understand their expectations, and manage their influence throughout the project lifecycle. A well-executed stakeholder analysis ensures better communication, stronger collaboration, and reduced risks of misalignment.

Traditionally, stakeholder analysis relies on manual research, interviews, and subjective assessments, which can be time-consuming and prone to biases. AI-powered tools are revolutionizing this process by automating stakeholder identification, analyzing sentiment, predicting engagement levels, and providing real-time recommendations.

This section explores how AI enhances stakeholder analysis, making it more data-driven, efficient, and insightful.

How AI Enhances Stakeholder Analysis

AI-powered stakeholder analysis tools leverage natural language processing (NLP), machine learning (ML), and predictive analytics to:

- **Identify key stakeholders** from multiple data sources

- **Assess their level of influence and interest** in the project

- **Analyze stakeholder sentiment** from emails, social media, and meeting transcripts

- **Predict potential areas of resistance or support**

- **Recommend communication strategies tailored to different stakeholders**

By integrating AI into stakeholder analysis, project managers can navigate complex stakeholder dynamics with greater precision and efficiency.

1. AI-Powered Stakeholder Identification
Finding Relevant Stakeholders

AI can scan multiple data sources – such as organizational charts, internal reports, meeting transcripts, email communications, and project documentation – to identify stakeholders who may impact or be impacted by the project.

For example, AI-powered platforms like Lucidchart AI and IBM Watson AI can:

- Automatically generate a list of stakeholders based on project scope

- Categorize stakeholders into primary (decision-makers), secondary (influencers), and tertiary (affected parties)

- Identify previously overlooked stakeholders who might play a critical role in project success

Mapping Stakeholder Influence and Interest

Once AI identifies stakeholders, it assesses their level of influence, interest, and engagement by analyzing past interactions, roles, and decision-making authority.

For example:

- A department head with approval authority may have high influence, high interest

- A junior employee affected by project changes may have low influence, high interest

- An executive sponsor overseeing the project may have high influence, moderate interest

AI-driven stakeholder mapping tools, such as Power BI and OrgMapper AI, can automatically generate influence-interest matrices, saving hours of manual analysis.

2. Sentiment Analysis for Stakeholder Engagement
AI-powered natural language processing (NLP) tools analyze stakeholder sentiment from emails, survey responses, social media, and meeting transcripts to gauge their level of support or resistance toward a project.

For example:

- If AI detects positive sentiment in stakeholder communications, it suggests opportunities to leverage their support.

- If AI detects negative sentiment (e.g., concerns about project risks, budget, or resource allocation), it flags these issues early, allowing project managers to proactively address concerns.

Example AI Tools for Sentiment Analysis:

- **IBM Watson NLP**: Analyzes stakeholder feedback from multiple sources to identify concerns and opportunities.

- **MonkeyLearn**: Uses AI to assess email and survey responses, categorizing them as positive, negative, or neutral.

- **Brandwatch**: Tracks sentiment from social media to understand how external stakeholders view the project.

This level of insight ensures that project managers can engage stakeholders effectively and address concerns before they escalate.

3. Predictive Analytics for Stakeholder Reactions
AI-driven predictive models analyze historical project data to forecast stakeholder behavior. This helps project managers anticipate potential resistance, conflicts, or advocacy before issues arise.

For example, AI can:

- Identify which stakeholders are likely to oppose a project based on past resistance to similar initiatives.

- Predict which stakeholders may require additional communication efforts to align with project goals.

- Recommend tailored engagement strategies to convert neutral stakeholders into active supporters.

Example AI Applications:

- **Tableau AI**: Uses predictive analytics to forecast stakeholder engagement trends based on historical interactions.

- **Crystal Knows**: Analyzes stakeholders' personalities and recommends the best communication styles for engagement.

By understanding how stakeholders are likely to react, project teams can adjust their strategies in advance, ensuring smoother collaboration.

4. AI-Generated Stakeholder Communication Plans
AI can automate and personalize communication strategies based on stakeholder preferences, engagement history, and sentiment analysis.

For instance, an AI tool can:

- **Segment stakeholders into communication groups** based on influence and interest

- **Recommend messaging strategies** (e.g., high-influence stakeholders require detailed reports, while low-interest stakeholders need brief updates)

- **Automate follow-up reminders** for key stakeholder check-ins

Example AI Communication Planning Tools:

- **Notion AI**: Automates meeting summaries and stakeholder updates.

- **Grammarly AI**: Enhances written communication to ensure clarity and professionalism.

- **Zoho CRM AI**: Personalizes email outreach and tracks engagement levels.

These tools help project managers maintain strong relationships with stakeholders, ensuring consistent alignment throughout the project lifecycle.

5. AI for Stakeholder Risk Management

AI-driven stakeholder analysis also plays a role in risk identification and mitigation. By continuously monitoring stakeholder interactions, AI can:

- Detect early warning signs of disengagement (e.g., decreasing email response rates, negative sentiment in meetings)

- Flag potential conflicts between stakeholders (e.g., differing opinions on project scope)

- Recommend mediation strategies based on previous successful resolutions

For example, an AI tool analyzing stakeholder emails might detect a drop in engagement from a key decision-maker. The system could then alert the project manager and suggest a proactive check-in meeting to re-engage the stakeholders.

This predictive approach helps project teams avoid surprises and navigate stakeholder challenges more effectively.

Creating Detailed Project Charters

A project charter serves as the foundational document that formally authorizes a project and outlines its objectives, scope, stakeholders, and key deliverables. It provides a clear direction for the project team and establishes alignment between stakeholders.

Traditionally, creating a project charter requires extensive manual effort involving stakeholder consultations, documentation reviews, and multiple iterations. AI-powered tools can streamline this process by automating document drafting, ensuring alignment with organizational goals, and enhancing clarity and consistency.

AI-Powered Drafting of Project Charters
AI can generate structured project charters based on minimal input, reducing the time and effort required for manual drafting. By analyzing past project data, industry standards, and organizational objectives, AI tools can suggest:

- **Project objectives and goals** based on historical success patterns

- **Scope definition** using best practices and similar project benchmarks

- **Stakeholder roles and responsibilities** derived from organizational structures

- **Risk identification and mitigation strategies** based on industry-specific risks

AI-driven platforms like Notion AI, ChatGPT, and Jasper AI allow project managers to input key details such as project name, industry, and expected outcomes. The system then generates a preliminary project charter, which can be refined and customized as needed.

For example, an AI-generated project charter for a software development initiative may automatically include functional and non-functional requirements, expected milestones, and compliance considerations.

Structuring Key Components with AI

AI tools help ensure that all essential sections of a project charter are included and formatted consistently. The key components typically generated by AI include:

Project Purpose and Justification

AI can analyze market trends, competitive intelligence, and internal business goals to provide a data-driven rationale for the project. Instead of relying solely on manual inputs, AI can extract relevant insights from past reports, business cases, and strategic plans to strengthen the project's justification.

Example AI Prompt: *"Generate a project justification for an AI-driven customer support chatbot for an e-commerce company, incorporating industry trends, customer expectations, and business benefits."*

Project Objectives and Success Criteria

AI can suggest SMART (Specific, Measurable, Achievable, Relevant, Time-bound) objectives tailored to the project scope. By analyzing similar projects and industry benchmarks, AI ensures that objectives are realistic and aligned with business priorities.

Example Output:

- **Objective:** Reduce customer support response time by 40% within six months.

- **Success Criteria:** Achieve an 80% customer satisfaction score for AI-driven responses.

Project Scope Definition

Scope definition is critical for preventing scope creep. AI tools can help outline what is included and excluded in the project scope, ensuring clarity from the start. AI-generated scope statements often include:

- **Deliverables** based on project goals

- **Boundaries and constraints** based on feasibility analysis

- **Assumptions and dependencies** derived from historical project data

Example Output: *"The chatbot project will focus on automating customer inquiries related to order tracking, refunds, and product availability. It will not handle complex issue resolution requiring human intervention."*

Stakeholder Roles and Responsibilities

AI can map stakeholders based on organizational structures and past project roles, ensuring that each key participant is clearly identified along with their responsibilities. This eliminates the ambiguity in team expectations.

Example Output:

- **Project Sponsor:** Oversees overall project alignment with business goals.

- **Product Owner:** Defines chatbot functionality and customer experience goals.

- **IT Team:** Implements chatbot infrastructure and integration with CRM.

Risk Identification and Mitigation

By leveraging AI-powered risk assessment tools, project managers can automatically identify common risks associated with similar projects and generate preemptive mitigation strategies. AI can assess historical project failures, industry challenges, and real-time data to enhance risk planning.

Example Risk and Mitigation Output:

- **Risk:** Chatbot may provide inaccurate responses, leading to customer frustration.

- **Mitigation Strategy:** Implement continuous AI model training and human oversight for complex queries.

Project Timeline and Milestones

AI-driven project management tools like Monday.com, Asana AI, and Microsoft Project AI can automatically generate project timelines and key milestones based on scope and complexity. These tools consider:

- Task dependencies

- Team availability

- Historical project durations

- Risk-adjusted timelines

Example AI-Generated Milestone Plan:

- **Week 1-4:** Define chatbot requirements and train the AI model

- **Week 5-8:** Develop initial chatbot prototype and test responses

- **Week 9-12:** Conduct user acceptance testing and refine the model

- **Week 13:** Deploy chatbot and monitor live performance

AI-Driven Consistency and Compliance Checks

Project charters must adhere to organizational standards, industry regulations, and compliance frameworks. AI-powered tools can analyze the document to:

- Detect missing components

- Ensure language clarity and consistency

- Check for regulatory compliance (e.g.. GDPR for data privacy, ISO standards for quality)

For example, a healthcare-related project charter could be reviewed by AI to ensure it includes HIPAA compliance measures[10] for handling patient data. Similarly, AI tools can flag vague or contradictory statements that need clarification before stakeholder approval.

AI-Powered Collaboration for Project Charter Development

AI enhances collaboration by enabling real-time updates and automated version control. Integrated AI tools within collaboration platforms like Google Docs AI, Confluence AI, and Slack AI help teams:

- Collect stakeholder feedback on project charters

- Suggest revisions based on real-time project updates

---

[10]https://www.forbes.com/sites/shashankagarwal/2023/12/22/ai-and-hipaa-navigating-the-privacy-crossroads/

- Track changes and ensure version control

For example, when a project scope change is requested, AI can automatically highlight the affected sections in the charter and suggest updated timelines, budgets, and resource allocations accordingly.

AI-Enabled Approvals and Stakeholder Buy-In
Once a project charter is finalized, securing stakeholder approval is critical. AI-driven platforms can streamline this process by:

- Automating stakeholder review workflows

- Summarizing key charter highlights for decision-makers

- Analyzing approval trends and stakeholder concerns based on past projects

If stakeholders express concerns, AI can generate alternative proposals or highlight sections that require further discussion. This ensures that project charters receive faster approvals with minimal delays.

AI Tools for Project Charter Creation
Notion AI and Jasper AI
- Automates drafting of structured project charters

- Ensures clarity and coherence in objectives and scope

IBM Watson AI for Compliance Checks
- Scans project charters for industry-specific compliance risks

- Identifies missing legal and regulatory elements

Lucidchart AI for Stakeholder Mapping
- Creates automated stakeholder influence diagrams

- Visualizes roles and responsibilities for alignment

Microsoft Project AI for Timeline Generation
- Automatically generates project schedules and dependencies

- Adjusts milestone plans based on risk assessments

Identifying Potential Project Risks and Mitigation Strategies

Risk management is an essential component of project planning. Unidentified or poorly managed risks can lead to budget overruns, missed deadlines, and project failure. Traditionally, risk identification and mitigation rely on manual assessments, past experiences, and subjective judgment. AI-powered tools now enable project managers to identify risks faster, more accurately, and with predictive insights, ensuring that mitigation strategies are data-driven and proactive.

This section explores how AI enhances risk identification and mitigation, the different types of project risks, and AI-driven solutions to minimize potential issues.

AI-Enhanced Risk Identification
AI-powered risk management tools use historical data, machine learning models, and real-time analytics to detect potential project risks before they escalate. These tools analyze past project failures, market conditions, resource constraints, and stakeholder behaviors to predict where problems may arise.

Key AI Capabilities for Risk Identification

- **Pattern Recognition:** AI scans past projects and identifies risk trends based on similarities in scope, budget, and execution.

- **Real-Time Monitoring:** AI continuously tracks project progress, detecting anomalies that indicate potential risks (e.g., delayed task completion, resource overuse).

- **Sentiment Analysis:** AI analyzes stakeholder communications (emails, meeting transcripts, feedback) to detect dissatisfaction or resistance that could evolve into project risks.

- **Automated Compliance Checks:** AI tools scan project plans for regulatory, legal, and industry compliance gaps, reducing risks of penalties or project delays.

Example AI Tools for Risk Identification

- **IBM OpenPages:** AI-powered risk assessment for regulatory compliance and financial risks.

- **RiskWatch AI:** Automates risk assessments and provides mitigation recommendations.

- **Microsoft Project AI:** Predicts schedule risks based on historical project performance.

Common Project Risks and AI-Driven Mitigation Strategies

AI enables project managers to not only identify risks early but also implement data-driven mitigation strategies. Below are some of the most common project risks and how AI can help mitigate them.

A. Scope Creep (Uncontrolled Expansion of Project Scope)

Risk: Unplanned scope changes lead to increased workload, higher costs, and delayed timelines.

AI-Driven Mitigation:

- AI-powered change management tools track scope modifications in real time and flag deviations from the original project plan.

- Predictive analysis determines the impact of additional requirements before approval.

- AI-based requirement validation tools ensure that new project demands align with business goals before being incorporated.

Example Tool:

- **Jira AI:** Detects unauthorized scope changes and assesses their impact.

B. Budget Overruns

Risk: Inaccurate cost estimates or unexpected expenses result in budget shortfalls.

AI-Driven Mitigation:

- AI analyzes historical cost data and improves budget forecasting accuracy.

- AI-driven resource allocation tools optimize spending by recommending cost-efficient strategies.

- AI-powered financial tracking tools flag excessive spending in real time, allowing early corrective action.

Example Tool:

- **Tableau AI:** Uses predictive analytics to monitor spending trends and detect financial risks.

C. Scheduling Delays

Risk: Poor time management or unexpected bottlenecks lead to project delays.

AI-Driven Mitigation:

- AI-powered scheduling tools analyze task dependencies and recommend optimized workflows.

- Machine learning models predict potential task delays based on team productivity patterns.

- AI-generated risk scenarios help project managers prioritize critical path tasks to minimize delays.

Example Tool:

- **Microsoft Project AI:** Automatically adjusts schedules based on workload analysis.

D. Resource Allocation Issues

Risk: Inefficient use of human and material resources leads to burnout, delays, and underutilization.

AI-Driven Mitigation:

- AI analyzes team availability and skill sets to recommend optimized workload distribution.

- AI-based task assignment tools ensure balanced workloads and prevent burnout.

- AI monitors resource utilization trends to prevent overstaffing or understaffing.

Example Tool:

- **ClickUp AI:** Uses predictive insights to balance workloads effectively.

E. Vendor and Supply Chain Disruptions

Risk: Third-party delays or failures disrupt project execution.

AI-Driven Mitigation:

- AI-powered supply chain analytics predict disruptions based on market trends and external factors.

- AI suggests alternative vendors or suppliers in case of delays.

- AI-driven contract risk assessment ensures vendor reliability before agreements are finalized.

Example Tool:

- **SAP AI:** Forecasts supplier risks and suggests contingency plans.

F. Stakeholder Conflicts

Risk: Misaligned stakeholder interests cause delays, resistance, or disruptions.

AI-Driven Mitigation:

- AI-powered sentiment analysis detects early signs of disagreement from stakeholder communications.

- AI recommends tailored engagement strategies to improve stakeholder alignment.

- AI tracks historical stakeholder behaviors to anticipate potential conflicts.

Example Tool:

- **Crystal Knows:** Analyzes stakeholder personality traits for improved communication strategies.

AI-Powered Risk Response Planning

AI not only helps identify and mitigate risks but also assists in developing proactive risk response plans. These plans ensure that projects recover quickly from setbacks with minimal disruption.

AI-Driven Risk Response Capabilities

- **Automated Risk Alerts:** AI sends real-time alerts when risk probability increases.

- **Scenario Simulations:** AI runs "what-if" analyses to evaluate potential responses to emerging risks.

- **Adaptive Risk Management:** AI continuously updates mitigation plans based on changing project conditions.

Example Tool:

- **RiskLens AI:** Provides automated risk impact assessments and response recommendations.

AI-Enabled Continuous Risk Monitoring

Traditional risk management follows a static approach, where risks are assessed periodically. AI enables continuous, real-time risk monitoring by integrating with project management systems and providing dynamic risk updates.

How AI Monitors Risks in Real-Time

- **Analyzing live project data** to detect anomalies.

- **Monitoring team productivity trends** to predict potential bottlenecks.

- **Detecting external risks** (e.g., regulatory changes, market shifts) that may impact project execution.

Example Tool:

- **IBM Watson AI for Risk Management:** Uses AI-driven insights to continuously monitor evolving project risks.

Prompt Frameworks for Comprehensive Requirements Documentation

Capturing and documenting project requirements is essential to ensuring that all stakeholders share a common understanding of project goals, scope, and deliverables. AI-powered tools can automate and enhance this process, but their effectiveness depends on how well the prompts are structured. A well-designed prompt extracts clear, concise, and actionable insights, helping project managers define accurate requirements.

This section provides a structured framework for AI prompts that assist in gathering, analyzing, and refining project requirements. These prompt templates can be used with AI tools like ChatGPT, Jasper AI, Notion AI, and business intelligence platforms to streamline documentation.

AI-Powered Requirement Gathering Prompts

AI can assist in generating functional, non-functional, business, and technical requirements by analyzing past projects, industry standards, and stakeholder input. The following prompts help structure requirement gathering efficiently.

A. Defining Functional Requirements

Functional requirements describe what the system, product, or service should do.

Prompt Example:

"Generate a list of functional requirements for a [software/product/service] that allows users to [perform specific actions]. Ensure that the requirements cover key features such as [authentication, data entry, user roles, integrations]."

Example Output for an E-commerce Website:

- Users must be able to create an account and log in securely.

- The system should allow customers to add items to a shopping cart and complete a purchase.

- An order confirmation email should be sent after successful transactions.

B. Defining Non-Functional Requirements

Non-functional requirements define how the system should perform (e.g., security, performance, usability).

Prompt Example:

"List the top non-functional requirements for a [industry] application that must handle [specific user load, security constraints, compliance needs]. Include aspects like performance, security, scalability, and usability."

Example Output for a Banking App:

- The system must handle 10,000 concurrent users without performance degradation.

- Data encryption must comply with PCI DSS security standards.

- The application should load within three seconds on mobile devices.

C. Gathering Business Requirements

Business requirements define the high-level needs and objectives of a project from a business perspective.

Prompt Example:

"Identify five key business requirements for a [project type] that aims to [achieve specific goals]. Ensure alignment with organizational objectives and customer expectations."

Example Output for a Customer Support Chatbot:

- Reduce customer response time by 50% within six months.

- Increase customer satisfaction ratings by 20% using AI-powered assistance.

- Ensure integration with existing CRM for personalized customer interactions.

D. Defining Technical Requirements

Technical requirements specify the underlying architecture, platforms, and technologies required for the project.

Prompt Example:

"List the technical requirements for a [software/system] that must integrate with [existing technology stack]. Include

programming languages, database choices, and API requirements."

Example Output for a Healthcare Management System:

- The system should use AWS cloud services for scalability.

- Data must be stored in a HIPAA-compliant encrypted database.

- APIs must support RESTful architecture for third-party integration.

AI-Powered Requirement Validation Prompts
Once requirements are gathered, AI can validate, refine, and detect inconsistencies to ensure completeness and feasibility.

A. Checking Requirement Completeness
Prompt Example:

"Review the following requirements for a [project type] and suggest any missing critical requirements. Ensure alignment with industry standards and best practices."

Example Output:

- Missing error-handling functionality in system requirements.

- No mention of mobile responsiveness for a web-based platform.

B. Ensuring Requirement Clarity and Precision
Prompt Example:

"Rewrite the following requirements to improve clarity, specificity, and measurability. Make sure each requirement follows the SMART criteria."

Example Input:

- "The system should support high traffic."

Example Output:

- "The system must handle 100,000 simultaneous user requests with a 99.9% uptime guarantee."

C. Identifying Conflicting Requirements
Prompt Example:

"Analyze the following set of requirements and identify any contradictions or conflicts. Suggest resolutions to ensure alignment."

Example Conflict:

- Requirement 1: "The system should allow guest checkout without requiring an email.'

- Requirement 2: "All orders must be linked to a unique customer email."

Suggested Resolution:

- "Allow guest checkout but require an email at the payment step for order tracking."

AI-Powered Stakeholder Requirement Analysis

Different stakeholders often have varying and sometimes conflicting expectations. AI can help analyze stakeholder input to ensure alignment.

A. Extracting Key Requirements from Stakeholder Feedback
Prompt Example:

"Summarize the main concerns and requirements from the following stakeholder meeting transcripts. Categorize them into functional, non-functional, business, and technical requirements."

B. Prioritizing Stakeholder Requirements
Prompt Example:

"Analyze the following stakeholder requirements and prioritize them based on business impact, feasibility, and urgency. Provide a ranking from 1 to 5."

Example Output for a Mobile App Project:

1. User authentication security enhancement (High priority – required for compliance).

2. Dark mode UI feature (Low priority – aesthetic preference).

3. Integration with third-party payment gateways (High priority – business-critical).

AI-Powered Risk Assessment in Requirement Gathering
Poorly defined requirements introduce risks. AI can analyze requirements to predict potential risks and recommend mitigations.

A. Identifying Requirement Gaps
Prompt Example:

"Analyze the following project requirements and identify potential gaps or missing elements that could cause project delays or failures."

B. Assessing Requirement Volatility
Prompt Example:

"Evaluate the likelihood of requirement changes for a [project type]. Suggest strategies to handle requirement volatility and avoid scope creep."

Example Output:

- High likelihood of regulatory changes → Implement modular architecture for adaptability.

- Feature creep risk due to multiple stakeholder inputs → Enforce change request process.

AI-Powered Requirement Documentation Templates
AI can automatically generate structured requirement documents based on collected inputs.

Prompt Example:

"Generate a full requirements document for a [project type], including sections for functional, non-functional, business, and technical requirements."

Example Output:

- Section 1: Introduction

- Section 2: Project Scope

- Section 3: Functional Requirements

- Section 4: Non-Functional Requirements

- Section 5: Business and Stakeholder Needs

- Section 6: Risk Assessment

AI tools like Notion AI, Google Docs AI, and Confluence AI help structure automated requirement documentation, reducing manual effort.

Chapter 6: Resource Allocation and Team Optimization

Efficient resource allocation and team optimization are critical to the success of any project. Assigning the right people to the right tasks, balancing workloads, and ensuring that teams operate at peak efficiency can be challenging – especially in complex projects with shifting priorities. Traditionally, resource management relies on manual planning, intuition, and historical performance data, which can lead to inefficiencies, misallocations, and burnout.

AI is transforming how organizations approach resource allocation by providing data-driven insights, predictive workload management, and intelligent team composition analysis. AI-powered tools analyze employee skills, past performance, and project needs to ensure that resources are allocated optimally. These technologies minimize inefficiencies, prevent team overload, and enhance productivity, ensuring that project goals are met effectively.

In this chapter, we will explore how AI enhances resource allocation and team optimization. By integrating AI into resource allocation and team management, organizations can achieve greater efficiency, higher employee satisfaction, and improved project outcomes.

AI-Powered Team Composition Analysis

Building the right team is fundamental to project success. A well-composed team ensures optimal collaboration, efficient problem-solving, and high productivity. However, traditional team composition methods often rely on manual selection, subjective judgment, and past experiences, which can lead to skill mismatches, imbalanced workloads, and underutilization of talent.

AI-powered team composition analysis leverages machine learning, predictive analytics, and collaboration data to form teams that are best suited for specific projects. AI tools analyze factors such as skills, experience, personality traits, collaboration history, and project requirements to recommend the most effective team structure. This results in stronger team dynamics, improved efficiency, and better project outcomes.

How AI Enhances Team Composition Analysis

AI-driven team analysis optimizes composition by considering multiple factors:

Skills and Expertise Matching

AI scans employee profiles, past project performance, and external qualifications to match the right individuals to project roles. It ensures that each role is filled with someone who possesses the necessary technical and soft skills to succeed.

Example: AI may recommend a full-stack developer with prior experience in fintech for a banking software project, rather than just any available developer.

AI Tools:

- **Recruitee AI** – Matches employees to projects based on skills and experience.

- **Workday AI** – Identifies skill gaps and suggests team members with relevant expertise.

Personality and Collaboration Analysis

Beyond technical skills, team chemistry plays a crucial role in performance. AI assesses collaboration styles, work habits, and past interactions to recommend individuals who work well together.

Example: If past data indicates that two designers collaborate effectively but struggle with communication when paired with a specific developer, AI can adjust team composition to prevent friction.

AI Tools:

- **Humanyze AI** – Analyzes team collaboration patterns.

- **Crystal Knows** – Provides personality insights for better team alignment.

Diversity and Inclusion Optimization

AI promotes diverse and inclusive teams by removing unconscious bias from team selection. It evaluates gender, cultural background, cognitive diversity, and work styles to create balanced teams that foster innovation and creativity.

Example: AI may suggest a mix of junior and senior team members to ensure mentorship opportunities while maintaining efficiency.

AI Tools:

- **Eightfold AI** – Ensures diversity in team selection.

- **Pymetrics** – Uses neuroscience-based AI to remove hiring and selection bias.

Project-Specific Adaptability

AI tailors team composition to the unique demands of each project by considering industry trends, required expertise, and projected challenges. If a project requires agility and fast decision-making, AI might recommend team members with prior experience in Agile methodologies.

Example: For a high-speed product launch, AI may select a team experienced in rapid iteration and lean development rather than those accustomed to longer waterfall-style projects.

AI-Driven Adjustments for Dynamic Projects

Team needs change throughout a project lifecycle. AI continuously monitors team performance and workload and suggests adjustments, such as:

- Adding a data analyst if the project shifts towards analytics-heavy tasks.

- Rotating team members to prevent burnout when workload increases.

- Replacing a team member who is underperforming or has scheduling conflicts.

This real-time adaptability ensures that teams remain high-performing throughout the project.

AI Tools:

- **ClickUp AI** – Tracks team workload and suggests reallocation.

- **Asana AI** – Recommends team structure changes based on shifting project needs.

By leveraging AI-powered team composition analysis, organizations optimize team performance, enhance collaboration, and ensure project success.

Skills Mapping and Resource Matching

Effective project execution depends on having the right people in the right roles. Skills mapping and resource matching ensure that each team member is assigned tasks that align with their expertise, experience, and strengths. However, traditional resource allocation methods often rely on manual tracking, outdated spreadsheets, and subjective judgment, leading to inefficiencies and skill mismatches.

AI-powered tools revolutionize this process by automating skill assessment, identifying resource gaps, and dynamically matching employees to project needs. AI can analyze vast amounts of data, including past project performance, employee career histories, learning patterns, and industry benchmarks, to optimize resource allocation.

How AI Enhances Skills Mapping

AI-driven skills mapping involves the systematic identification, categorization, and assessment of employee skills. This ensures that project teams are competent, well-balanced, and adaptable to project demands.

Automated Skill Identification

AI scans internal databases, employee resumes, past project contributions, performance reviews, and learning records to create a real-time skills inventory.

Example: AI can detect that an employee who completed a cybersecurity certification course and worked on network security projects is a strong candidate for an upcoming cybersecurity-related task.

AI Tools:

- **Workday AI** – Tracks employee skills and suggests career development opportunities.

- **LinkedIn Talent Insights** – Identifies workforce capabilities and emerging skill trends.

Predictive Skill Gap Analysis
AI analyzes current workforce skills and compares them against future project requirements, highlighting gaps that need to be addressed.

Example: If a company plans to implement AI-powered customer service, AI may flag a lack of employees trained in AI chatbot development and suggest either training programs or external hiring.

AI Tools:

- **Eightfold AI** – Predicts skill gaps and recommends hiring or upskilling solutions.

- **IBM Watson Talent Frameworks** – Maps employee capabilities against industry benchmarks.

Dynamic Skill Categorization
AI categorizes skills into three levels:

1. **Core Competencies** – Essential skills for a given role (e.g., coding for a software engineer).

2. **Adjacent Skills** – Related abilities that can be leveraged (e.g., UX design for a front-end developer).

3. **Emerging Skills** – New capabilities an employee is developing (e.g., AI-driven automation for an IT specialist).

By dynamically categorizing skills, AI ensures that employees can be matched to roles that utilize both their primary expertise and growing capabilities.

AI-Powered Resource Matching

Once skills are mapped, AI matches employees to tasks and projects based on:

- Project complexity and skill requirements

- Employee workload and availability

- Collaboration history and team compatibility

- Long-term career development goals

Intelligent Task Assignments

AI analyzes past project success rates and recommends the most suitable team members for a given task.

Example: If an AI system recognizes that **a** software engineer has successfully led past cloud migration projects, it may suggest assigning them to a similar upcoming initiative.

AI Tools:

- **Jira AI** – Matches engineers to software development tasks based on past performance.

- **ClickUp AI** – Suggests optimized workload distribution based on team strengths.

Balancing Workload to Prevent Burnout

AI monitors individual workloads to ensure that tasks are evenly distributed and prevent employee exhaustion.

Example: If an AI system detects that a team member is consistently working longer hours than peers, it can suggest redistributing tasks or recommending additional resources.

AI Tools:

- **Asana AI** – Adjusts task assignments based on real-time workload tracking.

- **Microsoft Viva Insights** – Detects burnout risk and recommends workflow adjustments.

Personalized Learning and Upskilling Suggestions
AI not only matches employees to existing skills but also suggests personalized learning paths to prepare them for future roles.

Example: If an AI tool identifies a growing demand for Python programming skills within an organization, it can recommend relevant training courses to employees with adjacent programming experience.

AI Tools:

- **Coursera AI for Business** – Suggests upskilling courses based on career goals.

- **LinkedIn Learning AI** – Recommends personalized learning paths for career growth.

By leveraging AI-driven skills mapping and resource matching, organizations ensure that employees are assigned to roles where they thrive, skill gaps are addressed proactively, and team efficiency is maximized.

Predictive Workload Management

Managing workloads effectively is critical to maintaining productivity, preventing burnout, and ensuring timely project completion. However, traditional workload management relies on static schedules, manual tracking, and reactive adjustments, which can lead to overburdened employees, resource underutilization, and missed deadlines. AI-powered predictive workload management leverages historical data, real-time analytics, and machine learning to forecast workload demands and optimize task distribution before issues arise.

AI tools can analyze team performance trends, predict bottlenecks, and dynamically adjust work assignments to balance workloads efficiently. This proactive approach ensures that project teams maintain a sustainable pace, reducing stress while maximizing efficiency.

How AI Enhances Workload Management

AI-powered workload management systems assess real-time project data, task complexity, employee capacity, and upcoming deadlines to optimize task distribution and prevent overloading any single team member.

A. Predicting Workload Peaks and Bottlenecks

AI analyzes past projects and current task progress to forecast when workload spikes are likely to occur. This allows managers to prepare for increased demand by adjusting deadlines, reallocating resources, or hiring temporary staff.

Example: AI may detect that a software development team is entering a high-risk phase of a project where multiple testing cycles overlap, triggering a warning for managers to redistribute testing tasks to additional team members.

AI Tools:

- **Microsoft Project AI** – Predicts workload spikes based on past trends.

- **Trello AI** – Identifies potential bottlenecks and recommends task redistribution.

B. Dynamic Workload Redistribution

AI continuously monitors task completion rates, employee availability, and project timelines to recommend real-time workload adjustments.

Example: If an AI system detects that one designer has completed 80% of their tasks ahead of schedule while another is struggling to meet deadlines, it may suggest reassigning upcoming work to maintain project balance.

AI Tools:

- **Asana AI** – Automatically reallocates tasks based on real-time project status.

- **Jira AI** – Adjusts developer workloads dynamically in Agile projects.

C. Employee Performance and Fatigue Monitoring

AI evaluates employee productivity patterns to detect early signs of burnout and recommend workload reductions when necessary.

Example: If AI identifies that an employee has been working overtime for multiple weeks, it may suggest delegating part of their workload or offering additional breaks to maintain long-term efficiency.

AI Tools:

- **Microsoft Viva Insights** – Tracks employee work habits and detects burnout risks.

- **Workday AI** – Provides recommendations for balanced task distribution.

D. Forecasting Resource Requirements

AI analyzes project scope, team capabilities, and task dependencies to predict whether additional resources will be needed before project deadlines.

Example: If AI predicts that a marketing campaign requires double the usual design workload due to a high volume of deliverables, it may recommend hiring a freelancer or reassigning tasks to avoid last-minute delays.

AI Tools:

- **ClickUp AI** – Forecasts future resource demands based on project scope.

- **Monday.com AI** – Analyzes workload trends and suggests additional staffing needs.

AI-Driven Scheduling Optimization

AI-powered scheduling tools help teams prioritize high-impact tasks, automate deadline adjustments, and ensure that no team member is overburdened or underutilized.

A. Intelligent Task Prioritization

AI evaluates task urgency, dependencies, and potential delays to recommend the optimal order of task completion.

Example: If AI identifies that delays in backend development will affect the front-end team, it may suggest reprioritizing tasks to prevent workflow disruptions.

AI Tools:

- **Smartsheet AI** – Prioritizes critical tasks to avoid project slowdowns.

- **Wrike AI** – Adjusts task sequences dynamically to optimize efficiency.

B. Automated Deadline Adjustments

AI predicts whether tasks will be completed on time based on historical performance and real-time data, making deadline adjustments when necessary.

Example: If AI detects that a documentation team is moving slower than expected due to additional client revisions, it may extend their deadline while adjusting later project milestones to maintain overall delivery timelines.

AI Tools:

- **Forecast AI** – Adjusts project schedules based on task progress.

- **Hive AI** – Suggests flexible deadlines when delays are detected.

Preventing Employee Overload While Maximizing Efficiency

Predictive workload management balances efficiency with wellbeing by ensuring that employees remain productive without being overwhelmed.

A. Custom Workload Balancing for Individual Employees

AI creates personalized workload profiles for each employee based on their working speed, capacity, and performance trends, ensuring fair task distribution.

Example: If AI recognizes that a developer consistently completes complex coding tasks faster than their peers, it may suggest adjusting their workload to maintain a sustainable pace while optimizing project output.

AI Tools:

- **Tempo AI** – Creates custom workload plans for each employee.

- **Teamwork AI** – Monitors individual work habits and adjusts assignments accordingly.

B. AI-Driven Productivity Enhancement Without Overworking Teams

AI suggests workflow improvements, automation opportunities, and collaboration strategies to help teams work smarter, not harder.

Example: AI may recommend automating repetitive report generation tasks to free up time for higher-value project activities.

AI Tools:

- **Notion AI** – Automates documentation and reduces repetitive workload.

- **Zapier AI** – Suggests workflow automation for increased efficiency.

By leveraging AI-powered predictive workload management, organizations ensure that tasks are evenly distributed, burnout is prevented, and projects stay on schedule without unnecessary strain on employees.

Performance Optimization Recommendations

Optimizing team performance is essential for ensuring that projects are completed on time, within budget, and at the highest quality. Traditional performance management relies on manual reviews, subjective feedback, and periodic assessments, which often fail to provide real-time insights or proactive improvements. AI-powered performance optimization leverages data-driven insights, machine learning algorithms, and real-time feedback to enhance individual and team productivity.

By analyzing work patterns, collaboration dynamics, task efficiency, and productivity trends, AI helps project managers and team leaders identify performance gaps, recommend targeted improvements, and create a culture of continuous enhancement.

How AI Enhances Performance Optimization

AI-powered performance optimization tools focus on individual efficiency, team collaboration, and overall project effectiveness by analyzing key productivity metrics and suggesting improvements.

A. AI-Driven Productivity Analysis

AI continuously monitors task completion rates, time spent on activities, and efficiency levels to identify patterns that affect productivity.

Example: AI may detect that team members spend excessive time on administrative tasks, suggesting automation to free up more time for strategic work.

AI Tools:

- **Microsoft Viva Insights** – Analyzes work habits and productivity trends.

- **RescueTime AI** – Tracks time usage and identifies productivity bottlenecks.

B. Personalized Performance Improvement Plans

AI helps create individualized development plans based on employees' strengths, weaknesses, and career goals.

Example: If AI identifies that a junior developer struggles with debugging complex code, it may suggest targeted training courses or mentorship from senior developers.

AI Tools:

- **Workday AI** – Recommends upskilling opportunities based on performance trends.

- **LinkedIn Learning AI** – Suggests personalized learning paths to improve skills.

C. Intelligent Task Delegation for Higher Efficiency

AI evaluates who is best suited for a specific task based on their skills, past performance, and workload capacity.

Example: If AI detects that a content writer consistently produces high-quality reports but struggles with graphic design, it may recommend assigning design-related tasks to another team member while maximizing the writer's focus on writing.

AI Tools:

- **Trello AI** – Suggests optimized task assignments based on historical performance.

- **ClickUp AI** – Distributes work based on skills and past efficiency levels.

D. AI-Powered Collaboration Optimization

AI enhances teamwork by analyzing communication patterns, collaboration efficiency, and feedback loops to improve interaction among team members.

Example: AI may find that developers and designers are frequently misaligned due to unclear task handoffs, prompting recommendations for improved documentation or automated task transitions.

AI Tools:

- **Slack AI** – Analyzes team communication efficiency.

- **Miro AI** – Optimizes collaborative workflows for remote teams.

Real-Time Performance Feedback and Adjustments
Traditional performance reviews are often delayed, providing feedback too late to be actionable. AI enables real-time performance monitoring and instant feedback, allowing employees to adjust their work habits proactively.

A. Continuous Feedback Loops
AI provides immediate insights into work progress, quality, and efficiency, allowing for on-the-go performance improvements.

Example: AI may notify a marketing team that their content engagement rates have dropped, prompting them to refine their strategy before the campaign fails.

AI Tools:

- **15Five AI** – Delivers continuous performance feedback for employees.

- **Betterworks AI** – Aligns employee goals with performance improvements.

B. Automated Performance Benchmarking
AI compares individual and team performance against industry benchmarks and historical data, helping managers set realistic goals and identify areas for improvement.

Example: AI may report that a development team completes features 20% slower than the industry average, suggesting workflow optimization strategies.

AI Tools:

- **Tableau AI** – Benchmarks performance metrics against industry trends.

- **Kissflow AI** – Tracks team performance and identifies areas for optimization.

AI-Powered Motivation and Engagement Strategies
AI doesn't just optimize work processes – it also enhances team motivation, engagement, and job satisfaction to improve long-term performance.

A. AI-Driven Recognition and Rewards
AI can identify high-performing employees and recommend recognition or incentives based on contribution levels.

Example: If AI detects that a team member consistently completes high-impact tasks ahead of schedule, it may suggest a recognition bonus or leadership opportunities.

AI Tools:

- **Bonusly AI** – Automates peer recognition and employee rewards.

- **Motivosity AI** – Suggests personalized incentives based on performance.

B. Preventing Employee Burnout and Work Fatigue
AI monitors work patterns and engagement levels, flagging signs of burnout before they affect performance.

Example: AI may notice a significant drop in productivity and increased overtime hours, suggesting mandatory breaks, workload adjustments, or wellness programs.

AI Tools:

- **Wellable AI** – Recommends wellness programs to maintain productivity.

- **Microsoft Viva Wellbeing** – Tracks employee wellbeing and suggests improvements.

By leveraging AI for performance optimization, teams can enhance productivity, improve collaboration, and maintain engagement, leading to better project outcomes.

Prompt Techniques for Team Dynamics Assessment

Effective team dynamics are essential for successful project execution. Strong collaboration, clear communication, and a balance of skills and personalities contribute to high-performing teams. However, assessing team dynamics has traditionally been a subjective process, relying on observation, feedback, and experience. AI-powered tools now enable data-driven analysis of team interactions, communication styles, and collaboration effectiveness.

AI-driven prompt techniques allow project managers to extract actionable insights from AI tools such as ChatGPT, Notion AI, and IBM Watson AI to analyze team cohesion, conflict resolution, engagement levels, and collaboration efficiency. These prompts help managers identify potential issues early, optimize team workflows, and improve overall team performance.

Using AI Prompts for Team Dynamics Assessment

AI prompts can be used to analyze various aspects of team dynamics, including communication effectiveness, conflict resolution, leadership influence, and overall team engagement. Below are structured prompts that project managers can use to enhance team performance.

A. Assessing Communication Effectiveness

Clear and effective communication is crucial for any team. AI can analyze email exchanges, meeting transcripts, and team chats to assess clarity, response times, and tone.

Prompt Example:

"Analyze the last [X] team meetings and internal messages to identify patterns in team communication. Highlight areas where miscommunication, delays, or lack of clarity are affecting collaboration."

AI-Generated Insights:

- Team members frequently interrupt each other during discussions, leading to misunderstandings.

- Critical project updates take too long to circulate, causing inefficiencies.

- Written communications lack clarity, requiring multiple follow-ups.

Solution: Implement structured meeting agendas and improve documentation practices.

AI Tools:

- **Slack AI** – Monitors and improves team communication effectiveness.

- **Otter.ai** – Transcribes and summarizes meetings to detect gaps in clarity.

B. Evaluating Team Collaboration and Engagement

AI can assess engagement by analyzing participation levels, task contributions, and meeting involvement.

Prompt Example:

"Analyze team engagement levels based on participation in discussions, responsiveness to project updates, and contribution to shared documents. Identify team members who may be disengaged or overwhelmed."

AI-Generated Insights:

- Some team members contribute significantly less in discussions.

- A few employees show high workloads but low engagement, signaling burnout.

- Certain roles lack clear responsibilities, leading to passive participation.

Solution: Reassign tasks, encourage more structured discussions, and ensure workload balance.

AI Tools:

- **Microsoft Viva Insights** – Tracks engagement and collaboration patterns.

- **Miro AI** – Enhances remote team engagement through visual collaboration analysis.

C. Detecting Potential Conflicts and Misalignment

AI can analyze sentiment and tone in emails, chat messages, and meeting transcripts to detect underlying conflicts or tension within teams.

Prompt Example:

"Analyze recent team interactions and identify any signs of tension, frustration, or communication breakdowns. Provide insights into possible sources of conflict and suggest mitigation strategies."

AI-Generated Insights:

- Certain team members frequently disagree in discussions without resolution.

- Specific phrases in written communication indicate frustration or resistance.

- There is an imbalance in decision-making influence, causing dissatisfaction.

Solution: Facilitate open discussions, address concerns directly, and improve team alignment.

AI Tools:

- **Humanyze AI** – Assesses workplace sentiment and interaction patterns.

- **Crystal Knows** – Analyzes personality dynamics for better conflict resolution.

D. Understanding Leadership Influence on Team Performance

AI can assess how leadership styles impact team motivation, productivity, and engagement.

Prompt Example:

"Evaluate how leadership communication style affects team morale and performance. Identify any gaps in leadership effectiveness and suggest improvements."

AI-Generated Insights:

- Leaders provide frequent updates but lack one-on-one engagement with team members.

- Employees feel hesitant to provide upward feedback.

- The decision-making process is unclear, leading to confusion in task execution.

Solution: Increase leadership transparency, encourage feedback loops, and improve mentorship strategies.

AI Tools:

- **15Five AI** – Tracks leadership impact on employee engagement.

- **Betterworks AI** – Analyzes leadership effectiveness in performance reviews.

E. Optimizing Remote and Hybrid Team Dynamics

With remote and hybrid work models becoming the norm, AI can assess how well-distributed teams are collaborating and communicating.

Prompt Example:

"Analyze remote team engagement, collaboration effectiveness, and response times compared to in-office teams. Provide recommendations to improve remote work dynamics."

AI-Generated Insights:

- Remote employees respond slower to non-essential emails but are highly efficient in focused work.

- In-office teams dominate discussions in hybrid meetings, reducing remote input.

- Virtual collaboration tools are underutilized, leading to information gaps.

Solution: Implement clear virtual meeting protocols, ensure equal participation, and use collaborative tools effectively.

AI Tools:

- **Zoom AI Companion** – Tracks participation in virtual meetings.

- **Slack AI** – Analyzes remote communication effectiveness.

AI-Powered Continuous Team Monitoring and Improvement

AI can provide ongoing real-time monitoring of team interactions and suggest adjustments before small issues escalate.

A. AI-Generated Team Performance Reports

AI generates weekly or monthly reports summarizing key trends in team dynamics, highlighting areas of strength and potential concerns.

Prompt Example:

"Generate a performance report summarizing team collaboration, engagement, and communication trends over the last month. Provide actionable recommendations for improving team efficiency."

B. AI-Driven Teamwork Prediction Models

AI can predict which teams are likely to experience performance issues based on historical trends and current project conditions.

Prompt Example:

"Based on current workload, team interactions, and sentiment analysis, predict potential productivity issues and team misalignment risks over the next [X] weeks. Suggest preemptive solutions."

AI Tools:

- **Tableau AI** – Forecasts team performance trends.
- **ClickUp AI** – Predicts team workload bottlenecks before they occur.

By leveraging AI-powered prompt techniques for team dynamics assessment, project managers can enhance communication, detect early signs of conflict, improve engagement, and create high-performing teams.

Chapter 7: Budget and Financial Management

Effective budget and financial management are critical to project success. Poor financial planning can lead to cost overruns, resource shortages, and project delays, ultimately impacting business profitability and stakeholder confidence. Traditionally, project budgeting and expense tracking rely on historical data, manual spreadsheets, and periodic reviews, which are prone to human error and reactive decision-making.

AI is transforming financial management by introducing predictive analytics, real-time expense monitoring, and automated cost optimization. AI-driven tools analyze historical spending patterns, market conditions, and project constraints to provide accurate budget forecasts, detect anomalies in expenditures, and assess financial risks before they escalate.

In this chapter, we will explore how AI enhances budget and financial management, including AI-driven budget forecasting, expense tracking and anomaly detection, financial risk assessment, cost optimization strategies, and prompts for financial modeling and analysis.

By integrating AI into financial management, organizations can increase budgeting accuracy, improve financial transparency, and make smarter cost-related decisions, ensuring projects stay within budget while maximizing value.

AI-Driven Budget Forecasting

Budget forecasting is one of the most critical aspects of financial management in project execution. Accurate forecasts help organizations allocate resources efficiently, prevent cost overruns, and ensure financial stability. However, traditional budget forecasting methods rely on historical data, manual calculations, and static financial models, which can lead to inaccuracies due to market fluctuations, unexpected expenses, or incomplete data.

AI-driven budget forecasting enhances accuracy and reliability by leveraging machine learning algorithms, predictive analytics, and real-time data integration. These technologies enable project managers and financial teams to anticipate budget needs, adjust financial plans dynamically, and minimize risks associated with financial uncertainty.

How AI Enhances Budget Forecasting
AI-driven budget forecasting improves financial planning by analyzing historical spending patterns, real-time economic indicators, and predictive cost trends. This results in more accurate, data-driven budget estimates that adapt to changing project conditions.

A. Predictive Cost Modeling
Traditional forecasting methods rely on past budget data and fixed assumptions. AI-powered predictive models analyze thousands of financial variables, including:

- Previous project costs and variances

- Market conditions and inflation rates

- Labor, material, and operational cost fluctuations

- Supplier pricing trends

126

Example: AI can predict that a construction project will require a 15% increase in raw material costs over six months based on supplier pricing trends and industry reports.

AI Tools:

- **IBM Planning Analytics** – Uses AI to predict budget requirements based on financial trends.

- **Anaplan AI** – Provides dynamic financial modeling for accurate cost projections.

B. Dynamic Real-Time Budget Adjustments

AI-powered forecasting continuously monitors actual expenditures and compares them with projected budgets. If deviations occur, AI suggests real-time adjustments to prevent financial mismanagement.

Example: If a software development project is exceeding its cloud hosting budget due to increased usage, AI may recommend switching to a cost-efficient provider or optimizing cloud resources.

AI Tools:

- **Adaptive Insights AI** – Tracks real-time budget variances and suggests corrective actions.

- **Oracle Financial Cloud** – Adjusts financial forecasts dynamically based on evolving cost trends.

C. Scenario Planning and What-If Analysis

AI enables scenario-based forecasting, allowing organizations to test multiple financial outcomes before making critical budget decisions.

Example: AI can simulate different financial scenarios for an expansion project, such as:

1. Best-case scenario – Market demand grows, increasing revenue by 20%.

2. Worst-case scenario – Supply chain disruptions increase costs by 10%.

3. Most likely scenario – A steady growth rate keeps the budget within limits.

AI Tools:

- **Tableau AI** – Runs financial simulations and visualizes scenario-based forecasts.

- **SAP Analytics Cloud** – Uses AI to model financial projections and risk-adjusted budgets.

D. AI-Powered Risk Forecasting in Budgeting

AI enhances financial risk assessment by predicting potential budget risks before they materialize. These risks include:

- **Scope creep** – Expanding project requirements leading to higher costs.

- **Vendor price increases** – Unexpected supplier cost hikes affecting budgets.

- **Regulatory changes** – Compliance costs due to new laws or industry standards.

Example: AI may detect that regulatory changes in data privacy laws will require additional cybersecurity investments, preventing unexpected compliance costs later.

AI Tools:

- **RiskLens AI** – Predicts financial risks and recommends cost-mitigation strategies.

- **Moody's Analytics AI** – Assesses economic risks that could impact project budgets.

AI-Driven Budget Forecasting in Different Industries

A. AI in Construction Budget Forecasting

- AI models predict raw material cost fluctuations and labor expenses.

- Forecasting tools optimize construction schedules to avoid costly delays.

Example: AI detects that a construction project will likely exceed its concrete budget by 8% due to supply chain disruptions, allowing managers to secure alternative suppliers in advance.

AI Tools:

- **Procore AI** – Predicts cost variations in construction projects.

- **BIM 360 AI** – Integrates budget forecasting with project schedules.

B. AI in IT and Software Development Budgeting

- AI predicts cloud service costs, developer salaries, and maintenance expenses.

- Tools recommend cost-effective software licenses and infrastructure solutions.

Example: AI forecasts that a software development team will exceed its budget by 12% due to overtime costs, prompting an early adjustment in project scope or deadlines.

AI Tools:

- **Jira AI** – Monitors software development costs in Agile projects.

- **AWS Cost Explorer AI** – Optimizes cloud computing expenses.

C. AI in Manufacturing and Supply Chain Finance

- AI predicts inventory costs, transportation expenses, and supply chain disruptions.

- Tools suggest cost-saving supplier contracts and demand forecasting adjustments.

Example: AI anticipates that shipping costs will rise by 15% due to fuel price increases, prompting supply chain managers to negotiate better freight contracts in advance.

AI Tools:

- **SAP Integrated Business Planning AI** – Forecasts supply chain costs and demand fluctuations.

- **Blue Yonder AI** – Predicts material shortages and suggests cost-saving solutions.

By integrating AI into budget forecasting, organizations can predict financial needs with greater accuracy, reduce cost overruns, and dynamically adjust budgets based on real-time insights.

Expense Tracking and Anomaly Detection

Managing project expenses effectively is essential for staying within budget and preventing financial waste. Traditional expense tracking methods rely on manual data entry, periodic financial reviews, and static reporting, which can lead to errors, delays, and undetected fraudulent activities. AI-powered expense tracking and anomaly detection revolutionize financial oversight by automating expense monitoring, flagging unusual transactions in real-time, and improving financial transparency.

With AI, organizations can detect spending irregularities, prevent budget leaks, and gain deeper insights into financial patterns, ensuring better financial control and compliance.

How AI Enhances Expense Tracking
AI-powered tools analyze historical spending patterns, invoice data, and real-time financial transactions to automate and improve expense tracking.

A. Real-Time Expense Monitoring
AI eliminates the need for manual expense reporting by continuously tracking financial transactions in real-time. It can:

- **Categorize expenses automatically** (e.g., travel, equipment, software).

- **Detect duplicate or unnecessary charges** that could inflate costs.

- **Provide instant alerts** when spending exceeds predefined limits.

Example: If AI detects an employee has submitted two identical expense claims, it automatically flags the duplication for review.

AI Tools:

- **Expensify AI** – Automates receipt tracking and categorizes expenses.

- **SAP Concur AI** – Provides real-time expense monitoring and fraud detection.

B. AI-Powered Receipt and Invoice Processing

AI-driven optical character recognition (OCR) extracts financial details from receipts and invoices, reducing manual data entry errors and improving efficiency.

Example: A consultant submits a photo of a travel receipt, and AI extracts key details (date, amount, and vendor) and categorizes the expense accordingly.

AI Tools:

- **Abacus AI** – Uses OCR to digitize and process receipts.

- **Rossum AI** – Automates invoice data extraction and validation.

AI in Anomaly Detection for Expense Fraud Prevention

Expense fraud and mismanagement can cost organizations millions in lost revenue. AI-powered anomaly detection algorithms identify suspicious transactions by analyzing:

- **Spending behaviors** that deviate from past trends.

- **Unusual vendor activity** (e.g., sudden spikes in payments).

- **Duplicate invoices or overinflated reimbursements**.

A. Identifying Unusual Spending Patterns

AI compares current expenses with historical trends to identify anomalies.

Example: If an employee submits a hotel expense 50% higher than usual for a location, AI may flag the discrepancy for manual review.

AI Tools:

- **Visa Smarter Payments AI** – Detects irregular spending in corporate accounts.

- **Fyle AI** – Flags duplicate and inflated expenses for finance teams.

B. Detecting Unauthorized Vendor Transactions

AI can cross-check vendor invoices against approved contracts to ensure accuracy.

Example: If AI identifies that a vendor has billed twice for the same service, it automatically notifies the finance team.

AI Tools:

- **Xero AI** – Detects unauthorized payments and fraudulent vendor transactions.

- **Anodot AI** – Provides anomaly detection in financial transactions.

C. Preventing Internal Financial Misconduct

AI-driven analytics monitor employee spending behaviors and highlight potential misuse of funds.

Example: AI detects that an employee is booking non-work-related flights under corporate travel expenses.

AI Tools:

- **AppZen AI** – Uses machine learning to audit expenses for compliance violations.

- **Coupa AI** – Monitors procurement and employee expenses for suspicious activity.

AI-Driven Expense Optimization and Cost Control
AI not only detects anomalies but also recommends cost-saving strategies by identifying inefficiencies in spending.

A. Identifying Redundant Subscriptions and Services
AI can track recurring payments for underutilized software, memberships, or services and suggest cancellations.

Example: AI notices that a company is paying for five different cloud storage solutions when only two are actively used.

AI Tools:

- **Gartner Expense AI** – Identifies redundant software expenses.

- **Zylo AI** – Tracks SaaS spending and optimizes subscription costs.

B. Predicting Future Cost Increases
AI forecasts rising vendor prices, inflation trends, or new regulatory fees that may impact budgets.

Example: AI alerts a manufacturing company about an upcoming 10% increase in raw material costs, allowing them to negotiate contracts early.

AI Tools:

- **Oracle NetSuite AI** – Predicts financial risks in supply chain spending.

- **Moody's Analytics AI** – Forecasts cost fluctuations and economic risks.

Best Practices for Implementing AI in Expense Tracking

1. **Integrate AI with Financial Systems** – Ensure AI tools connect with ERP, accounting, and procurement software for real-time data flow.

2. **Set Custom AI Rules** – Define spending limits, vendor rules, and fraud detection parameters for accurate monitoring.

3. **Conduct Regular AI Audits** – Validate AI findings to refine accuracy and fraud detection capabilities.

4. **Ensure Compliance with Financial Regulations** – AI tools should align with SOX, GDPR, and industry-specific financial regulations

By integrating AI into expense tracking and anomaly detection, organizations can reduce financial waste, prevent fraud, and optimize cost efficiency.

Financial Risk Assessment

Every project carries financial risks that can impact budgets, profitability, and overall business stability. These risks stem from market fluctuations, unexpected expenses, regulatory changes, supplier disruptions, and internal mismanagement. Traditionally, financial risk assessment relies on historical data analysis, periodic audits, and expert judgment, which can be slow, reactive, and prone to human bias.

AI-driven financial risk assessment introduces predictive analytics, real-time data monitoring, and automated risk scoring to help organizations identify, quantify, and mitigate financial risks before they escalate. AI enables continuous risk monitoring, ensuring that financial decisions are based on up-to-date insights rather than static, backward-looking reports.

How AI Enhances Financial Risk Assessment

AI-powered financial risk assessment improves decision-making by:

- **Detecting financial vulnerabilities early** through real-time monitoring.

- **Predicting potential risks** based on historical patterns and market trends.

- **Providing automated risk scoring** to prioritize mitigation efforts.

A. AI-Powered Predictive Risk Modeling

AI identifies financial risks before they become critical by analyzing historical financial performance, spending trends, and external market conditions.

Example: AI forecasts that a logistics company may experience a 15% cost increase due to rising fuel prices, allowing the company to adjust its transportation strategy proactively.

AI Tools:

- **SAP Risk Management AI** – Analyzes financial data to predict budget overruns.

- **Moody's Analytics AI** – Forecasts macroeconomic risks and market trends.

B. Real-Time Financial Risk Monitoring

Unlike traditional assessments that rely on periodic reviews, AI provides continuous risk monitoring by scanning financial transactions, contracts, and regulatory updates.

Example: AI detects that a supplier is experiencing financial distress, alerting the procurement team to secure an alternative vendor before disruptions occur.

AI Tools:

- **RiskLens AI** – Monitors financial and operational risks in real-time.

- **IBM OpenPages AI** – Tracks compliance risks and regulatory changes affecting budgets.

C. AI-Based Risk Scoring and Prioritization

AI assigns risk scores to different financial threats based on likelihood, severity, and potential business impact. This helps project managers prioritize risk mitigation strategies efficiently.

Example: AI assesses the risk impact of supply chain disruptions, inflation, and delayed client payments, ranking them from high to low priority.

AI Tools:

- **Oracle Risk Management AI** – Automates financial risk scoring and reporting.

- **FICO Falcon AI** – Detects and ranks fraud risks in financial transactions.

D. Identifying Internal Financial Risks

AI helps organizations detect internal mismanagement, policy violations, and potential fraud by analyzing financial transactions and behavioral patterns.

Example: AI flags an unusual pattern of small, repeated invoice adjustments, prompting a review for potential expense fraud.

AI Tools:

- **AppZen AI** – Audits financial transactions for fraud and compliance violations.

- **Workday Adaptive Planning AI** – Detects inconsistencies in budgeting and financial reports.

AI in Industry-Specific Financial Risk Assessment
A. AI in Banking and Finance

- AI predicts loan defaults and credit risks using customer transaction patterns.

- Detects suspicious financial activity in real time to prevent fraud.

Example: AI flags a series of high-value transfers from an account, triggering a fraud investigation.

AI Tools:

- **Darktrace AI** – Identifies real-time cyber and financial fraud threats.

- **FICO AI** – Monitors and scores credit risk assessments.

B. AI in Supply Chain and Manufacturing

- AI forecasts material price fluctuations and prevents overstocking or shortages.

- Assesses supplier reliability based on past performance and financial stability.

Example: AI alerts a manufacturing firm to potential supply shortages due to geopolitical instability.

AI Tools:

- **SAP Predictive Analytics AI** – Optimizes supply chain cost management.

- **Blue Yonder AI** – Predicts vendor-related financial risks.

C. AI in IT and Software Development
- AI tracks cloud service expenses and detects cost inefficiencies.

- Predicts contractual risks in software licensing and vendor agreements.

Example: AI suggests switching cloud providers to avoid a 20% cost increase due to new pricing models.

AI Tools:

- **AWS Cost Explorer AI** – Forecasts cloud computing expenses.

- **Jira AI** – Identifies software project budget risks.

By leveraging AI for financial risk assessment, organizations gain predictive insights, prevent budget disruptions, and ensure financial stability.

Cost Optimization Strategies

Controlling project costs while maintaining efficiency and quality is a constant challenge for organizations. Traditional cost management methods focus on budget constraints, periodic expense reviews, and manual cost-cutting decisions, which often lead to delayed responses, inefficiencies, and missed opportunities for savings. AI-driven cost optimization introduces predictive analytics, real-time expense monitoring, and automated recommendations, enabling businesses to reduce unnecessary expenses, maximize resource utilization, and improve financial sustainability.

By integrating AI-powered cost optimization strategies, organizations can make data-driven financial decisions, ensuring projects remain profitable, efficient, and within budget without compromising quality.

How AI Enhances Cost Optimization
AI-driven cost optimization helps organizations:

- **Analyze cost trends** to identify inefficiencies.

- **Predict future spending** and suggest preemptive cost-saving actions.

- **Automate cost allocation and budget adjustments** for maximum efficiency.

A. AI-Powered Cost Reduction through Data Analysis
AI analyzes historical financial data, operational costs, and market trends to identify areas where cost savings can be achieved without affecting performance.

Example: AI identifies excessive software licensing costs by tracking underutilized applications and recommends consolidating subscriptions to save money.

AI Tools:

- **SAP S/4HANA AI** – Analyzes operational spending for inefficiencies.

- **Tableau AI** – Provides insights into cost-saving opportunities based on data trends.

B. Optimizing Resource Utilization

AI ensures that human resources, technology, and materials are allocated efficiently to prevent under-utilization or overuse.

Example: AI detects that certain employees have lighter workloads and recommends redistributing tasks instead of hiring additional staff.

AI Tools:

- **Microsoft Project AI** – Allocates team resources based on workload predictions.

- **ClickUp AI** – Balances workloads to maximize productivity without unnecessary hiring.

C. Dynamic Vendor and Supplier Cost Optimization

AI-driven procurement tools help negotiate better pricing, compare vendor costs, and prevent supply chain overpayments.

Example: AI flags a 20% cost increase from a supplier and suggests switching to an alternative vendor offering the same quality at a lower price.

AI Tools:

- **SAP Ariba AI** – Analyzes vendor pricing trends and cost efficiency.

- **GEP SMART AI** – Automates supplier negotiations for better contract terms.

AI-Driven Cost Avoidance Strategies

Instead of reactively cutting costs, AI helps organizations avoid unnecessary expenses before they occur.

A. Predictive Maintenance to Reduce Operational Costs

AI predicts equipment failures and maintenance needs, preventing expensive repairs and downtime.

Example: AI forecasts a 30% risk of a manufacturing machine failure within three months and recommends preventive maintenance.

AI Tools:

- **IBM Maximo AI** – Predicts maintenance needs to reduce repair costs.

- **Uptake AI** – Optimizes asset usage and extends equipment lifespan.

B. Reducing Cloud and IT Infrastructure Costs

AI optimizes cloud resource allocation, storage, and computing power to eliminate unnecessary IT expenses.

Example: AI identifies that servers are running at only 40% capacity overnight and recommends adjusting cloud usage to reduce costs.

AI Tools:

- **AWS Cost Explorer AI** – Suggests cost-saving strategies for cloud services.

- **Google Cloud AI** – Analyzes cloud consumption patterns to eliminate waste.

C. Automating Cost Control Policies

AI enforces cost control by setting automated spending limits, tracking budget deviations, and flagging excessive expenditures in real-time.

Example: AI detects excessive travel expenses in a specific department and suggests policy adjustments to curb unnecessary spending.

AI Tools:

- **Oracle Cloud Financials AI** – Monitors financial transactions for excessive spending.

- **Expensify AI** – Automates expense reporting with cost-control measures.

AI in Industry-Specific Cost Optimization

A. AI in Manufacturing and Supply Chain Management
- Optimizes inventory levels to reduce storage costs.

- Reduces waste and excess materials through predictive demand analysis.

Example: AI predicts lower seasonal demand for a product, preventing overproduction and unnecessary inventory costs.

AI Tools:

- **Blue Yonder AI** – Forecasts supply chain demand and prevents excess stock.

- **Llamasoft AI** – Optimizes logistics to reduce transportation expenses.

B. AI in Retail and E-Commerce
- Adjusts pricing strategies to maximize revenue while cutting costs.

- Identifies low-performing products that should be discontinued or optimized.

Example: AI detects that a particular product has declining sales and suggests discounting or discontinuation to reduce storage costs.

AI Tools:

- **Revionics AI** – Recommends dynamic pricing adjustments to optimize revenue.

- **Dynamic Yield AI** – Personalizes pricing and promotions to maximize sales.

C. AI in IT and Software Development

- Reduces software development costs by optimizing project timelines.

- Prevents unnecessary infrastructure spending in cloud services.

Example: AI suggests code refactoring to improve efficiency, reducing cloud processing costs for a large-scale AI application.

AI Tools:

- **Jira AI** – Analyzes software development costs and suggests optimizations.

- **New Relic AI** – Monitors cloud performance and prevents unnecessary costs.

By leveraging AI for cost optimization, organizations can reduce financial waste, maximize efficiency, and ensure long-term cost sustainability.

Prompts for Financial Modeling and Analysis

Financial modeling and analysis are essential for making informed decisions about budgeting, cost projections, investment strategies, and financial risk assessment. Traditionally, financial modeling requires manual data entry, static spreadsheets, and time-consuming calculations, which can lead to inefficiencies and potential errors. AI-powered tools enhance financial modeling by automating calculations, generating real-time insights, and predicting financial outcomes with greater accuracy.

To maximize the potential of AI in financial decision-making, project managers and financial analysts can use structured AI prompts to extract meaningful insights from AI-driven financial tools such as ChatGPT, IBM Watson AI, Oracle Financials AI, and Tableau AI. These prompts help generate budget forecasts, cost analysis, financial risk assessments, and investment recommendations tailored to specific business needs.

AI-Powered Prompts for Budget Forecasting
AI can assist in generating detailed financial forecasts by analyzing historical spending, market trends, and projected expenses.

A. Predicting Future Budget Needs
Prompt Example:

"Analyze the last [X] years of financial data for [company/project] and predict the budget requirements for the next [Y] years. Include expected cost variations due to inflation, economic trends, and project growth."

AI Output:

- Forecasted budget allocation across departments.

- Expected cost increases for materials, labor, and overhead.

145

- Adjustments based on economic and industry-specific factors.

AI Tools:

- **Anaplan AI** – Generates dynamic budget forecasts.

- **IBM Planning Analytics AI** – Predicts financial trends based on historical data.

B. Scenario-Based Budget Forecasting
Prompt Example:

"Create best-case, worst-case, and most-likely budget scenarios for [project name] over the next [X] years. Consider external risks such as supply chain disruptions, regulatory changes, and market volatility."

AI Output:

- Three scenario models with cost projections.

- Recommendations for budget contingency planning.

AI Tools:

- **Tableau AI** – Visualizes financial forecast scenarios.

- **Oracle Financial Cloud AI** – Models risk-adjusted financial scenarios.

AI Prompts for Expense Analysis and Cost Control
AI can identify cost inefficiencies, flag unnecessary expenditures, and recommend cost-saving measures.

A. Identifying Budget Overruns
Prompt Example:

"Analyze financial reports for the past [X] months and identify areas where spending has exceeded budget allocations. Provide recommendations for cost reduction."

AI Output:

- Expense categories with the highest overruns.

- Suggested spending cuts without affecting project performance.

AI Tools:

- **Expensify AI** – Tracks and categorizes project expenses.

- **Coupa AI** – Provides automated cost-saving recommendations.

Detecting Unnecessary or Redundant Expenses
Prompt Example:

"Identify recurring expenses or underutilized resources in [department/project] that can be reduced or eliminated. Suggest cost optimization strategies."

AI Output:

- Subscription services that are not actively used.

- Office resources that can be consolidated.

AI Tools:

- **Zylo AI** – Tracks and optimizes software subscription expenses.

- **SAP Ariba AI** – Analyzes procurement costs and suggests reductions.

AI Prompts for Financial Risk Assessment

AI helps identify financial risks associated with market fluctuations, project delays, or unexpected cost increases.

A. Assessing Financial Risks in a Project
Prompt Example:

"Analyze financial risks for [project name] over the next [X] years. Identify high-risk areas and suggest risk mitigation strategies."

AI Output:

- List of financial vulnerabilities.

- Recommended contingency planning measures.

AI Tools:

- **RiskLens AI** – Quantifies financial risk exposure.

- **FICO AI** – Detects fraud and compliance risks.

B. Predicting Cash Flow Challenges
Prompt Example:

"Evaluate cash flow trends for [company/project] over the past [X] quarters and predict potential liquidity challenges for the next [Y] quarters. Suggest corrective measures."

AI Output:

- Forecasted cash flow fluctuations.

- Strategies for improving liquidity management.

AI Tools:

- **Oracle Risk Management AI** – Monitors financial stability indicators.

- **Microsoft Dynamics AI** – Forecasts cash flow disruptions.

AI Prompts for Investment and Profitability Analysis
AI can assist businesses in making data-driven investment decisions by analyzing potential returns, market trends, and risk factors.

A. Evaluating Investment Opportunities
Prompt Example:

"Analyze potential return on investment (ROI) for [investment opportunity] over the next [X] years. Consider economic conditions, industry trends, and historical performance."

AI Output:

- ROI projections under different economic scenarios.

- Risk assessment and recommendations.

AI Tools:

- **Bloomberg Terminal AI** – Analyzes investment data and market trends.

- **Morningstar AI** – Provides AI-driven investment recommendations.

B. Profitability Optimization for an Existing Business
Prompt Example:

"Assess the profitability of [business unit/product] and recommend strategies to improve revenue while reducing operational costs."

AI Output:

- Breakdown of revenue-generating activities.

- Cost-cutting strategies without impacting productivity.

AI Tools:

- **Oracle Cloud Financials AI** – Optimizes profitability strategies.

- **Tableau AI** – Identifies cost-revenue imbalances.

AI Prompts for Financial Compliance and Regulatory Analysis

AI ensures financial decisions comply with regulatory requirements, tax laws, and industry standards.

A. Ensuring Regulatory Compliance
Prompt Example:

"Analyze financial transactions for [company/project] over the past [X] months to identify any compliance risks related to [specific regulation]. Suggest corrective actions."

AI Output:

- List of transactions flagged for potential compliance issues.

- Suggested policy changes to avoid regulatory penalties.

AI Tools:

- **IBM OpenPages AI** – Automates financial compliance analysis.

- **AppZen AI** – Audits expenses for policy violations.

150

B. Tax Optimization Strategies
Prompt Example:

"Evaluate tax liabilities for [company] and suggest legal tax optimization strategies for the next fiscal year."

AI Output:

- Identified deductions and credits.

- Tax-saving recommendations.

AI Tools:

- **TurboTax AI** – Analyzes tax-saving opportunities.

- **Xero AI** – Optimizes tax planning for businesses.

By leveraging AI-powered prompts for financial modeling and analysis, organizations can improve forecasting accuracy, detect cost inefficiencies, optimize investments, and mitigate financial risks.

Chapter 8: Timeline and Schedule Management

Effective timeline and schedule management is essential for ensuring projects are completed on time, within scope, and without unnecessary delays. Traditional scheduling methods often rely on manual planning, fixed Gantt charts, and historical estimates, which can lead to rigid timelines, inefficient resource allocation, and unexpected bottlenecks. As projects become more complex, organizations need smarter, more adaptive scheduling solutions.

AI-powered scheduling tools enhance project timeline management by automating task prioritization, optimizing critical paths, and predicting potential delays before they happen. By leveraging machine learning, real-time data analysis, and historical performance insights, AI helps project managers create flexible, optimized schedules that adapt to changing circumstances.

In this chapter, we will explore how AI enhances timeline and schedule management, including:

- **Intelligent project scheduling** – Automating task sequencing and deadline setting.

- **Critical path optimization** – Identifying the most efficient way to complete a project.

- **Predictive timeline adjustments** – Using AI to anticipate and mitigate delays.

- **Resource leveling using AI** – Ensuring balanced workloads across teams.

- **Prompt templates for schedule creation and monitoring** – Extracting AI-driven insights for real-time adjustments.

By integrating AI-driven scheduling techniques, project managers can increase efficiency, minimize risks, and ensure timely project completion in dynamic work environments.

Intelligent Project Scheduling

Project scheduling is a complex process that involves defining tasks, estimating durations, setting dependencies, and allocating resources effectively. Traditional scheduling methods rely on manual planning, historical estimates, and fixed deadlines, which can lead to inefficiencies, misaligned priorities, and unforeseen delays. AI-powered scheduling transforms this process by automating task sequencing, analyzing constraints, and dynamically adjusting schedules in real-time.

AI-driven project scheduling tools use machine learning, predictive analytics, and historical project data to create adaptive, optimized schedules that improve project efficiency and reduce scheduling conflicts.

How AI Enhances Project Scheduling

AI-powered scheduling systems optimize project timelines by:

- **Automating task assignments** based on project scope and priorities.

- **Analyzing dependencies and constraints** to prevent scheduling conflicts.

- **Adjusting schedules dynamically** based on real-time progress and resource availability.

A. AI-Generated Task Sequencing and Prioritization

AI evaluates task dependencies, deadlines, and available resources to determine the most efficient order of execution.

Example: AI identifies that Task A must be completed before Task B, but Task C can run in parallel, optimizing workflow efficiency.

AI Tools:

- **Microsoft Project AI** – Suggests optimized task sequences based on priority and dependencies.

- **Jira Advanced Roadmaps AI** – Automatically generates Agile sprint schedules.

B. Real-Time Schedule Adjustments

AI-powered scheduling tools continuously monitor progress, resource constraints, and task delays to make dynamic adjustments.

Example: If a key developer is unexpectedly unavailable, AI reassigns tasks to available team members or suggests timeline extensions to prevent project disruption.

AI Tools:

- **ClickUp AI** – Adjusts project timelines based on real-time team capacity.

- **Smartsheet AI** – Identifies schedule risks and recommends adjustments.

C. Forecasting Delays and Optimizing Deadlines
AI analyzes past project performance and identifies patterns that could lead to delays, allowing managers to take proactive action.

Example: AI predicts that similar projects have experienced delays in the testing phase, prompting the team to allocate additional testing resources upfront.

AI Tools:

- **Oracle Primavera AI** – Predicts scheduling risks in large-scale projects.

- **Monday.com AI** – Uses machine learning to optimize deadlines based on past performance.

AI-powered project scheduling reduces manual effort, enhances flexibility, and ensures that project timelines remain realistic and achievable.

Critical Path Optimization

The critical path in project management represents the longest sequence of dependent tasks that determine the shortest possible project duration. Any delay in these tasks will directly impact the overall project completion date. Traditional critical path analysis requires manual calculations, static scheduling techniques, and historical estimates, which can lead to inefficiencies, miscalculations, and rigid timelines.

AI-powered tools optimize the critical path by automating task sequencing, analyzing dependencies, and predicting potential bottlenecks. These AI-driven solutions ensure that projects run on the most efficient timeline, dynamically adjusting schedules when needed to prevent delays.

How AI Enhances Critical Path Optimization

AI-driven critical path analysis improves scheduling by:

- **Identifying high-risk tasks** that could cause delays.

- **Optimizing dependencies** to reduce task bottlenecks.

- **Providing real-time adjustments** to maintain project momentum.

A. AI-Powered Critical Path Identification

AI evaluates all project tasks and automatically determines the sequence of critical activities that will define project duration.

Example: AI identifies that **Task A, Task D, and Task G** are the longest dependencies in a construction project, marking them as **critical** and suggesting resource allocation adjustments to prevent slowdowns.

AI Tools:

- **Oracle Primavera AI** – Identifies critical tasks in complex projects.

- **Microsoft Project AI** – Analyzes task dependencies and adjusts the critical path accordingly.

B. Predicting and Preventing Critical Path Delays

AI uses historical project data and real-time tracking to predict which critical tasks are at risk of delays and recommends proactive solutions.

Example: AI detects that Task C has historically taken 20% longer than estimated and suggests adding buffer time or extra resources to mitigate potential delays.

AI Tools:

- **Smartsheet AI** – Uses historical insights to adjust scheduling buffers.

- **ClickUp AI** – Notifies managers when critical tasks are falling behind schedule.

C. Dynamic Critical Path Adjustments

AI-powered scheduling tools automatically recalculate the critical path when tasks are delayed or completed ahead of time.

Example: If AI recognizes that a design phase was completed early, it adjusts downstream tasks to advance the project schedule accordingly.

AI Tools:

- **Jira Advanced Roadmaps AI** – Adjusts Agile sprint schedules dynamically.

- **Monday.com AI** – Reoptimizes critical task sequences in real-time.

AI-driven critical path optimization ensures that projects are completed in the shortest possible time while mitigating risks and maintaining flexibility.

Predictive Timeline Adjustments

Project timelines are often disrupted by unexpected delays, resource constraints, and changing priorities. Traditional scheduling methods rely on fixed deadlines and manual adjustments, which can lead to reactive decision-making and inefficiencies. AI-powered predictive timeline adjustments enhance project management by using real-time data, historical trends, and machine learning to forecast potential delays and suggest proactive scheduling changes.

AI-driven timeline management ensures greater flexibility, reduced risks, and improved efficiency by continuously analyzing project progress and dynamically adjusting schedules to maintain realistic deadlines.

How AI Enhances Predictive Timeline Adjustments

AI tools monitor project progress, predict scheduling risks, and recommend real-time adjustments to keep projects on track.

A. AI-Driven Delay Prediction

AI analyzes historical project data, current task completion rates, and resource availability to forecast potential delays before they occur.

Example: AI detects that previous software development projects faced testing delays, prompting a recommendation to extend testing timelines or allocate additional QA resources.

AI Tools:

- **Microsoft Project AI** – Predicts delays based on task completion trends.

- **Oracle Primavera AI** – Forecasts scheduling risks in large-scale projects.

B. Automated Schedule Adjustments

AI dynamically reorganizes task sequences, reallocates resources, and adjusts deadlines in response to real-time changes.

Example: If a key developer calls in sick, AI identifies the best available team member to take over and suggests shifting non-critical tasks to balance workloads.

AI Tools:

- **ClickUp AI** – Reallocates tasks based on team availability.

- **Smartsheet AI** – Automatically adjusts project deadlines when delays are detected.

C. Optimizing Timeline Buffers

AI helps managers set strategic buffer times by analyzing past project durations and identifying where extra time is needed.

Example: AI recognizes that design approval processes tend to take longer than planned and recommends adding a 10% buffer to design review stages.

AI Tools:

- **Jira Advanced Roadmaps AI** – Adjusts Agile sprint timelines dynamically.

- **Monday.com AI** – Suggests optimal buffer times based on past performance.

AI-powered predictive timeline adjustments ensure that project schedules remain flexible, minimizing risks while maintaining efficiency.

Resource Leveling Using AI

Effective resource management is crucial for maintaining project efficiency and preventing burnout. Traditional resource-leveling methods involve manually adjusting workloads, redistributing tasks, and balancing team availability, which can be time-consuming, prone to errors, and reactive rather than proactive. AI-powered resource leveling automates this process by analyzing workloads, optimizing team assignments, and ensuring balanced resource distribution across project phases.

By leveraging AI for resource leveling, project managers can avoid overloading key employees, minimize downtime, and ensure optimal use of available resources, leading to improved productivity and reduced project risks.

How AI Enhances Resource Leveling

AI-driven resource leveling focuses on predicting workload imbalances, reallocating tasks dynamically, and preventing resource shortages or underutilization.

A. AI-Powered Workload Balancing

AI continuously monitors team capacity, task complexity, and deadlines to distribute workloads evenly across employees.

Example: AI detects that one developer has a significantly heavier workload than others and automatically redistributes some tasks to maintain balance.

AI Tools:

- **Microsoft Project AI** – Identifies workload imbalances and reallocates tasks accordingly.

- **ClickUp AI** – Ensures even distribution of work among team members.

B. Forecasting Resource Bottlenecks

AI predicts potential resource shortages or excess capacity by analyzing past project data and current assignments.

160

Example: AI identifies that a critical project phase will require more design resources than available, prompting managers to adjust hiring plans or reassign tasks in advance.

AI Tools:

- **Smartsheet AI** – Forecasts resource availability issues and suggests solutions.

- **Jira Advanced Roadmaps AI** – Predicts sprint workload imbalances for Agile teams.

C. Real-Time Resource Adjustments
AI dynamically adjusts resource allocations based on task progress, shifting priorities, and unexpected changes.

Example: If AI detects that a team member is out sick for a week, it automatically suggests reallocating their tasks to available colleagues without affecting deadlines.

AI Tools:

- **Oracle Primavera AI** – Reallocates resources dynamically in complex projects.

- **Monday.com AI** – Provides real-time workload optimization recommendations.

AI-powered resource leveling ensures that teams remain productive without being overworked, improving efficiency and maintaining a steady workflow.

Prompt Templates for Schedule Creation and Monitoring

AI-powered scheduling tools can automate schedule creation, monitor progress in real-time, and predict potential delays, but their effectiveness depends on how well they are utilized. To maximize AI's potential, project managers need well-structured prompts that extract relevant insights, automate adjustments, and provide actionable recommendations.

This section provides prompt templates that can be used with AI tools like ChatGPT, Microsoft Project AI, Smartsheet AI, and Jira Advanced Roadmaps AI to streamline project scheduling, monitor progress, and optimize timelines dynamically.

AI Prompts for Initial Schedule Creation

AI can assist in generating optimized project schedules by analyzing task dependencies, resource availability, and deadlines.

A. Creating a Project Timeline from Scratch
Prompt Example:

"Generate a detailed project timeline for [project name] with key milestones, dependencies, and estimated task durations. Use industry best practices to ensure an efficient schedule."

AI Output:

- Task breakdown with estimated durations.

- Critical dependencies and sequential task ordering.

- Recommended milestone tracking points.

AI Tools:

- **Microsoft Project AI** – Generates structured project timelines.

- **Monday.com AI** – Creates automated task sequences.

B. Optimizing Task Dependencies
Prompt Example:

"Analyze the dependencies in the following project schedule and recommend optimizations to reduce bottlenecks and shorten the timeline."

AI Output:

- Identification of inefficient dependencies.

- Recommendations for parallel task execution.

AI Tools:

- **Jira Advanced Roadmaps AI** – Suggests Agile sprint adjustments.

- **Oracle Primavera AI** – Identifies dependency conflicts in complex schedules.

AI Prompts for Critical Path Analysis
The critical path determines the minimum time required to complete a project. AI can identify high-priority tasks and potential delays.

A. Identifying the Critical Path
Prompt Example:

"Determine the critical path for [project name] based on the current task list and dependencies. Highlight the tasks that have no flexibility and may impact the final deadline."

AI Output:

- List of tasks on the critical path.

- Task sequence recommendations for faster completion.

AI Tools:

- **Smartsheet AI** – Visualizes the project's critical path.

- **ClickUp AI** – Provides real-time adjustments for critical tasks.

B. Mitigating Critical Path Delays
Prompt Example:

"Analyze our project's critical path and recommend strategies to reduce completion time without compromising quality."

AI Output:

- Suggestions for parallel task execution.

- Additional resource allocation recommendations.

AI Tools:

- **Microsoft Project AI** – Recalculates project timelines based on delay risks.

- **Jira AI** – Suggests Agile sprint optimizations.

AI Prompts for Predictive Timeline Adjustments
AI can predict schedule risks and recommend real-time changes based on evolving project conditions.

A. Forecasting Schedule Risks
Prompt Example:

"Analyze our project progress and identify any tasks at risk of delay. Provide recommendations for adjusting the timeline."

AI Output:

- Tasks with high delay risks.

- Recommended schedule changes to maintain deadlines.

AI Tools:

- **Oracle Primavera AI** – Predicts delays in large-scale projects.

- **Monday.com AI** – Notifies teams of potential scheduling risks.

B. Dynamic Timeline Adjustments
Prompt Example:

"Given the following project status updates, adjust the schedule to reflect changes in task completion rates and resource availability."

AI Output:

- Updated project timeline with new task deadlines.

- Adjusted resource allocations to meet deadlines.

AI Tools:

- **Smartsheet AI** – Auto-updates schedules based on real-time input.

- **ClickUp AI** – Rebalances workloads dynamically.

AI Prompts for Resource Leveling and Workload Balancing

AI ensures that workloads remain balanced and that resources are allocated efficiently.

A. Detecting Resource Overload
Prompt Example:

"Analyze the current resource allocation and identify any team members who are overburdened. Suggest workload distribution improvements."

AI Output:

- Overloaded team members and their tasks.

- Recommendations for task redistribution.

AI Tools:

- **Microsoft Project AI** – Identifies and resolves workload imbalances.

- **Jira Advanced Roadmaps AI** – Reassigns tasks dynamically for Agile teams.

B. Optimizing Resource Assignments
Prompt Example:

"Based on available resources, suggest an optimized task distribution to improve efficiency and prevent burnout."

AI Output:

- Task assignments with balanced workloads.

- Recommended shifts in resource allocation.

AI Tools:

- **Smartsheet AI** – Provides real-time workload balancing.

- **Oracle Primavera AI** – Adjusts staffing for large-scale projects.

AI Prompts for Real-Time Schedule Monitoring
AI continuously monitors progress, alerts teams to potential issues, and suggests corrective actions.

A. *Monitoring Project Progress*
Prompt Example:

"Provide a summary of the current project status, highlighting completed tasks, ongoing activities, and tasks at risk of delay."

AI Output:

- Project progress percentage.

- Status of each task and flagged delays.

AI Tools:

- **ClickUp AI** – Generates real-time progress updates.

- **Microsoft Project AI** – Tracks project performance.

B. *Generating Automated Schedule Reports*
Prompt Example:

"Generate a project schedule report including key milestones, upcoming deadlines, and any deviations from the original plan."

AI Output:

- Overview of key project deadlines.

- Identification of on-track vs. delayed tasks.

AI Tools:

- **Tableau AI** – Visualizes project data in interactive reports.

- **Monday.com AI** – Summarizes project schedules for leadership teams.

By using AI-powered prompts for schedule creation and monitoring, project managers can automate scheduling, predict delays, and optimize project timelines dynamically.

Chapter 9: Communication and Collaboration

Effective communication and collaboration are at the heart of successful project management. Without clear messaging, timely updates, and seamless teamwork, even the most well-planned projects can face delays, misunderstandings, and inefficiencies. Traditionally, project teams rely on emails, meetings, and manual reporting, which can lead to information overload, misalignment, and wasted time. AI is transforming the way teams communicate and collaborate by automating routine updates, summarizing key discussions, and ensuring that the right information reaches the right people at the right time.

AI-powered tools enhance communication by analyzing language, summarizing meetings, automating reports, and optimizing team collaboration. Whether it's generating concise status updates, identifying communication bottlenecks, or facilitating cross-team coordination, AI ensures that project teams stay aligned, informed, and productive.

This chapter explores how AI-driven communication strategies streamline workflows, reduce manual effort, and improve team synergy. It covers AI-enhanced communication strategies, automated status reporting, intelligent meeting summarization, and cross-team collaboration optimization. Additionally, it provides prompt frameworks for effective communication, enabling project managers to extract clear, actionable insights from AI-powered assistants.

By integrating AI into project communication, teams can reduce friction, improve transparency, and foster a more efficient, connected work environment.

AI-Enhanced Communication Strategies

Effective communication is critical to project success, ensuring that teams remain aligned, stakeholders stay informed, and workflows progress smoothly. However, traditional communication methods—such as lengthy email chains, inconsistent messaging, and manually compiled reports—can lead to miscommunication, delays, and inefficiencies. AI-powered communication strategies address these challenges by automating, analyzing, and optimizing how information is shared within project teams.

AI-driven communication tools use natural language processing (NLP), sentiment analysis, and machine learning to enhance clarity, improve response times, and ensure that key messages reach the right audience. These tools help streamline updates, reduce manual effort, and foster better collaboration by eliminating redundant communication and providing real-time, data-driven insights.

How AI Enhances Communication in Project Management

AI-powered communication tools improve efficiency by:

- **Automating routine messaging** and eliminating unnecessary back-and-forth emails.

- **Improving clarity and tone** in written communication.

- **Analyzing sentiment** to detect and resolve potential miscommunication issues.

- **Prioritizing important messages** and summarizing key takeaways.

A. AI-Generated Real-Time Updates

AI automates project updates by extracting data from task management systems and generating concise, actionable reports.

Example: Instead of manually drafting progress updates, AI tools can scan project data and automatically send daily or weekly status reports to stakeholders.

AI Tools:

- **Microsoft Teams AI** – Generates automated status updates from integrated project tools.

- **ClickUp AI** – Summarizes team activity and pending tasks for easier decision-making.

B. Sentiment Analysis for Team Communication

AI can analyze emails, chat messages, and feedback to detect frustration, confusion, or potential conflicts in communication.

Example: If AI detects negative sentiment in team discussions about a delayed milestone, it can alert managers to address concerns proactively.

AI Tools:

- **IBM Watson Tone Analyzer** – Evaluates sentiment in emails and messages.

- **Crystal Knows** – Analyzes communication styles to improve team interactions.

C. AI-Powered Email and Message Optimization

AI improves clarity, tone, and professionalism in written communication by suggesting edits, rewording unclear messages, and removing unnecessary jargon.

Example: AI can refine an email draft by making it more concise, improving the tone, and suggesting a call to action for better engagement.

AI Tools:

- **Grammarly AI** – Enhances clarity and tone in professional communication.

- **ChatGPT AI** – Assists in drafting effective emails and responses.

By leveraging AI-enhanced communication strategies, project teams can reduce miscommunication, enhance collaboration, and improve overall efficiency.

Automated Status Reporting

Regular status reporting is essential for tracking project progress, identifying risks, and keeping stakeholders informed. However, traditional status reporting relies on manual data collection, spreadsheet updates, and time-consuming report writing, which can lead to delays, inconsistencies, and outdated information. AI-powered automated status reporting transforms this process by generating real-time, data-driven reports that provide accurate updates without the need for manual intervention.

AI-driven reporting tools pull information from task management platforms, time-tracking systems, and communication channels, automatically compiling key insights into structured reports. This ensures that teams, managers, and stakeholders receive timely and relevant updates, enabling faster decision-making and reducing administrative workload.

How AI Enhances Status Reporting

AI-powered status reporting improves efficiency by:

- **Automating data extraction** from multiple sources.

- **Generating structured summaries** based on real-time project updates.

- **Customizing reports** for different stakeholders.

- **Providing predictive insights** to highlight risks before they escalate.

A. Real-Time Data Extraction and Report Generation

AI scans task completion rates, project milestones, and resource allocation to create real-time progress updates.

Example: Instead of manually compiling weekly reports, AI automatically pulls data from project management tools and generates an email summary for stakeholders.

AI Tools:

- **Microsoft Power BI AI** – Extracts data from multiple sources and visualizes reports.

- **ClickUp AI** – Generates automated task and milestone summaries.

B. Customizable AI-Generated Reports

AI tailors reports based on different audiences, ensuring that executives receive high-level summaries, while project teams get detailed task breakdowns.

Example: AI generates a one-page executive summary for leadership while providing a detailed action list for team leads.

AI Tools:

- **Monday.com AI** – Customizes status reports based on audience needs.

- **Notion AI** – Summarizes key project updates in an easy-to-read format.

C. Predictive Risk Identification in Reports
AI highlights potential risks in reports by analyzing project delays, bottlenecks, and workload imbalances.

Example: AI detects that multiple tasks are behind schedule and includes a risk alert in the report, recommending adjustments.

AI Tools:

- **Smartsheet AI** – Flags potential scheduling issues in reports.

- **Tableau AI** – Uses predictive analytics to identify upcoming risks.

By leveraging AI for automated status reporting, organizations can save time, improve reporting accuracy, and ensure stakeholders have real-time visibility into project progress.

Intelligent Meeting Summarization

Meetings are essential for project coordination, but they can also be time-consuming, unstructured, and overwhelming—especially when key takeaways are buried in long discussions. Traditionally, summarizing meetings requires manual note-taking and post-meeting documentation, which can lead to missed details, inconsistencies, and delays in action items. AI-powered meeting summarization tools address these challenges by automating transcription, extracting key insights, and generating concise summaries that help teams stay informed and aligned.

AI-driven summarization tools use natural language processing (NLP) and machine learning algorithms to analyze conversations, identify action items, and deliver structured summaries. This ensures that project teams can quickly review decisions, track responsibilities, and streamline follow-ups without having to revisit lengthy recordings or written notes.

How AI Enhances Meeting Summarization
AI-powered tools improve meeting efficiency by:

- **Automatically transcribing conversations** with high accuracy.

- **Extracting key discussion points** and action items.

- **Summarizing decisions** in a structured format.

- **Highlighting areas requiring follow-up or further clarification.**

A. AI-Powered Meeting Transcription
AI automatically converts spoken words into text, eliminating the need for manual note-taking.

Example: AI transcribes a project update meeting and categorizes discussions into agenda items, key decisions, and unresolved issues.

AI Tools:

- **Otter.ai** – Provides real-time meeting transcription and keyword tagging.

- **Fireflies.ai** – Generates searchable transcripts and integrates them with project tools.

B. AI-Generated Action Items and Key Takeaways
AI extracts critical information from discussions, including decisions made, next steps, and task assignments.

Example: AI summarizes a sprint planning meeting by listing:

1. Feature A must be tested by Friday (Assigned: QA Team).

2. Marketing materials for Product B launch need review (Owner: Sarah).

3. Budget adjustments are required before procurement approval (Finance Team).

AI Tools:

- **Microsoft Teams AI** – Auto-generates meeting notes and action items.

- **Notion AI** – Summarizes discussions into structured project updates.

C. AI-Powered Follow-Up Reminders
AI can schedule reminders and follow-ups based on meeting outcomes, ensuring that decisions lead to actionable results.

Example: AI detects that a decision needs approval by next week and automatically schedules a reminder for the responsible team.

AI Tools:

- **Fathom AI** – Generates follow-up emails with meeting highlights.

- **Zoom AI Companion** – Summarizes virtual meetings and tracks follow-up tasks.

By integrating AI-powered meeting summarization, project teams can save time, reduce miscommunication, and ensure follow-through on key decisions.

Cross-Team Collaboration Optimization

Collaboration between different teams is essential for project success, but coordinating across departments, locations, and workflows can be challenging. Misalignment, siloed information, and inefficient communication often lead to delays, duplicated work, and mismanaged resources. Traditionally, cross-team collaboration relies on manual coordination, shared documents, and scheduled meetings, which can be time-consuming and prone to miscommunication.

AI-powered tools transform cross-team collaboration by automating information sharing, aligning team priorities, and optimizing workflows. Using machine learning, natural language processing (NLP), and predictive analytics, AI ensures that all teams remain connected, informed, and working towards common objectives—even in complex, multi-department projects.

How AI Enhances Cross-Team Collaboration
AI-driven collaboration tools improve efficiency by:

- **Breaking down communication silos** and centralizing shared information.

- **Automating workflow synchronization** across departments.

- **Enhancing knowledge sharing** to avoid redundant work.

- **Detecting bottlenecks** and suggesting process improvements.

A. AI-Powered Knowledge Management and Information Sharing

AI ensures that relevant information is shared across teams in real-time, preventing knowledge gaps.

Example: AI detects that the marketing team needs product specifications and automatically pulls relevant details from the product development team's documents.

AI Tools:

- **Notion AI** – Organizes and recommends shared resources for different teams.

- **Guru AI** – Centralizes company knowledge and provides instant answers to team queries.

B. Intelligent Workflow Coordination and Automation

AI streamlines project workflows by automatically assigning tasks, tracking dependencies, and aligning team schedules.

Example: AI recognizes that the content team is waiting for legal approval on marketing materials and notifies the legal team to prioritize the review.

AI Tools:

- **Asana AI** – Automates task dependencies and cross-team assignments.

- **Zapier AI** – Integrates different apps to automate multi-team workflows.

C. Predictive Analytics for Collaboration Efficiency

AI analyzes team performance and workload balance to recommend adjustments that enhance productivity and reduce friction.

Example: AI flags that two teams are working on overlapping initiatives and suggests merging efforts to avoid redundancy.

AI Tools:

- **Microsoft Viva AI** – Monitors team collaboration trends and suggests improvements.

- **Trello AI** – Recommends task distribution to balance workloads across teams.

By leveraging AI for cross-team collaboration optimization, organizations can improve efficiency, reduce misalignment, and foster a seamless workflow between departments.

Prompt Frameworks for Effective Communication

Clear and effective communication is essential for project success, ensuring that teams stay aligned, stakeholders remain informed, and workflows proceed smoothly. AI-powered tools can enhance communication by automating messages, summarizing conversations, and generating structured responses based on well-crafted prompts. The effectiveness of these AI-driven communication tools depends on how prompts are structured to extract useful, concise, and actionable insights.

This section provides prompt frameworks for various communication needs, including status updates, stakeholder

179

engagement, conflict resolution, meeting follow-ups, and cross-team collaboration. These prompts can be used with AI-powered assistants such as ChatGPT, Microsoft Copilot, Notion AI, and Slack AI to streamline messaging and ensure clarity.

AI Prompts for Status Updates
AI can generate real-time status reports by extracting data from project management systems and summarizing key progress points.

A. Daily or Weekly Team Updates
Prompt Example:

"Generate a [daily/weekly] project update for [project name] summarizing completed tasks, ongoing activities, and upcoming milestones. Keep it concise and action-oriented."

AI Output:

- **Completed:** Task A, Task B

- **In Progress:** Task C, Task D

- **Next Steps:** Task E is due by Friday

AI Tools:

- **Microsoft Teams AI** – Sends automated project status messages.

- **ClickUp AI** – Compiles real-time task updates into a structured report.

B. Executive Summary Reports
Prompt Example:

"Summarize project progress for executive stakeholders, focusing on key achievements, risks, and upcoming deadlines in under 200 words."

AI Output:

- **Project Phase:** Development 85% complete

- **Risks:** Testing delays due to resource constraints

- **Next Steps:** Final testing scheduled for next week

AI Tools:

- **Monday.com AI** – Customizes executive reports.

- **Tableau AI** – Generates high-level data insights.

AI Prompts for Stakeholder Communication
AI can personalize messages for different stakeholders based on project progress, risks, or milestone achievements.

A. Informing Stakeholders About Delays
Prompt Example:

"Draft an email update to stakeholders explaining that [project name] is experiencing a delay due to [reason]. Include revised timelines and mitigation strategies."

AI Output:

"Due to unforeseen resource constraints, [project name] will be delayed by two weeks. We are reallocating additional resources to minimize further impact and expect to meet the revised deadline of [new date]."

AI Tools:

- **Grammarly AI** – Refines stakeholder messaging for clarity.

- **ChatGPT AI** – Generates structured stakeholder updates.

B. Communicating Milestone Achievements
Prompt Example:

"Generate a celebratory email for stakeholders announcing the completion of [milestone]. Highlight key achievements and next steps."

AI Output:

"We're excited to share that [milestone] has been successfully completed! This marks a major step forward in [project name], bringing us closer to our final goal. Thank you for your continued support!"

AI Tools:

- **Microsoft Copilot AI** – Creates polished milestone emails.

- **Notion AI** – Summarizes achievements for reporting.

AI Prompts for Conflict Resolution and Team Alignment
AI can analyze sentiment in team discussions and assist in conflict resolution by crafting diplomatic messages.

A. Addressing Miscommunication in a Team
Prompt Example:

"Draft a message to resolve a miscommunication between [team A] and [team B] regarding [specific issue]. Ensure a neutral and solution-oriented tone."

AI Output:

"It looks like there has been some misunderstanding regarding [issue]. Let's clarify expectations and ensure alignment moving forward. [Action items] have been set to streamline our approach."

AI Tools:

- **IBM Watson Tone Analyzer** – Detects sentiment and refines messaging.

- **Crystal Knows AI** – Personalizes messages based on communication styles.

B. Encouraging Cross-Team Collaboration
Prompt Example:

"Create a message encouraging better collaboration between [Team A] and [Team B] for [project name]. Emphasize shared goals and efficiency."

AI Output:

"To ensure a seamless workflow, we encourage close collaboration between [Team A] and [Team B]. By aligning our efforts, we can enhance efficiency and deliver the best results for [project name]."

AI Tools:

- **Slack AI** – Automates cross-team announcements.

- **Asana AI** – Facilitates workflow coordination.

AI Prompts for Meeting Summaries and Follow-Ups
AI can automatically transcribe, summarize, and extract key takeaways from meetings, reducing manual note-taking.

A. Summarizing a Meeting Discussion
Prompt Example:

"Summarize the key points from the following meeting transcript, highlighting action items and assigned responsibilities."

AI Output:

- **Decision:** Product launch moved to Q2
- **Action Items:**
    - Update marketing timeline (Assigned: Sarah)
    - Finalize testing phase (Assigned: Dev Team)

AI Tools:

- **Otter.ai** – Transcribes and summarizes meetings.
- **Fireflies.ai** – Extracts action items from discussions.

B. Generating Follow-Up Emails
Prompt Example:

"Draft a follow-up email summarizing today's meeting, including key decisions, pending tasks, and next steps."

AI Output:

"Thank you for today's productive meeting. Key takeaways include [summary]. Please ensure that action items are completed by [deadline]. Let me know if any clarifications are needed."

AI Tools:

- **Zoom AI Companion** – Automates meeting follow-up emails.

- **Notion AI** – Structures meeting summaries into action lists.

AI Prompts for Continuous Communication Monitoring

AI can continuously monitor team communication channels to detect potential issues and suggest improvements.

A. Detecting Communication Bottlenecks
Prompt Example:

"Analyze team communication patterns and identify any bottlenecks affecting collaboration. Provide recommendations for improvement."

AI Output:

- Slow response times on Slack impacting sprint progress.

- Weekly syncs could be replaced with shorter daily stand-ups.

AI Tools:

- **Microsoft Viva AI** – Evaluates communication trends.

- **Humanyze AI** – Analyzes team collaboration efficiency.

By leveraging AI-powered prompts for effective communication, reporting, and collaboration, project teams can improve clarity, enhance efficiency, and ensure seamless information flow across all levels of an organization.

Chapter 10: Performance Tracking and Continuous Improvement

Successful project management goes beyond meeting deadlines and staying within budget—it requires continuous evaluation and improvement. Traditional performance tracking methods rely on manual reporting, periodic reviews, and static key performance indicators (KPIs), which can lead to delayed insights and reactive decision-making. AI-powered performance tracking introduces real-time monitoring, predictive analytics, and automated reporting, allowing project managers to make data-driven decisions and proactively address risks.

AI enhances performance tracking by analyzing project health, predicting potential roadblocks, and continuously refining workflows. Instead of waiting for end-of-phase reviews, AI tools provide real-time insights into team efficiency, budget utilization, and task completion rates, ensuring projects stay on course.

This chapter explores AI-powered KPI monitoring, real-time project health assessments, and predictive performance analytics, offering practical ways to leverage AI for ongoing project evaluation and continuous improvement. Additionally, we will provide AI-driven prompt techniques to help project managers extract meaningful insights, identify performance gaps, and implement proactive adjustments.

By integrating AI into performance tracking, organizations can move from a reactive to a proactive approach, ensuring greater efficiency, reduced risks, and sustained project success.

AI-Powered KPI Monitoring

Key Performance Indicators (KPIs) are essential for measuring project success, tracking progress, and identifying areas for improvement. However, traditional KPI monitoring relies on manual data collection, static reports, and periodic reviews, which can lead to delayed insights, outdated metrics, and missed opportunities for proactive adjustments. AI-powered KPI monitoring transforms this process by automating data collection, analyzing trends in real-time, and providing predictive insights to keep projects on track.

AI-driven KPI monitoring tools integrate with project management software, financial systems, and resource-tracking platforms to continuously evaluate performance metrics. This ensures that project managers have up-to-date, actionable insights to make informed decisions and optimize workflows.

How AI Enhances KPI Monitoring
AI-powered tools improve KPI tracking by:

- **Automating data collection** from multiple sources.

- **Providing real-time updates** instead of waiting for manual reports.

- **Detecting performance trends and anomalies** before they become critical.

- **Predicting future performance** based on historical data.

A. Real-time KPI Tracking and Visualization

AI continuously updates KPIs by pulling data from project management and collaboration tools, ensuring teams have access to real-time performance insights.

Example: AI tracks project milestones, team productivity, and budget spending, displaying progress on a live dashboard for immediate review.

AI Tools:

- **Tableau AI** – Visualizes real-time KPI performance data.

- **Power BI AI** – Monitors project metrics and generates automated reports.

B. AI-Driven Performance Anomaly Detection

AI identifies deviations from expected KPI benchmarks, helping project managers detect potential issues early.

Example: AI flags that task completion rates have dropped by 15% compared to previous sprints, prompting a deeper investigation into workload distribution.

AI Tools:

- **IBM Watson AI** – Analyzes KPIs and detects deviations in performance trends.

- **ClickUp AI** – Alerts teams when productivity KPIs fall below thresholds.

C. Predictive KPI Analytics for Future Performance

AI doesn't just track current KPIs—it predicts future trends and risks based on historical data and ongoing performance.

Example: AI forecasts that a project is likely to exceed budget within the next two months based on current spending patterns.

AI Tools:

- **Smartsheet AI** – Predicts timeline and resource-related KPI fluctuations.

- **Monday.com AI** – Provides proactive performance recommendations.

By leveraging AI-powered KPI monitoring, project teams can eliminate manual tracking, gain real-time insights, and take proactive measures to optimize project success.

Real-Time Project Health Assessment

Assessing project health is critical to ensuring that tasks, resources, budgets, and timelines remain on track. Traditional project health assessments rely on periodic reviews, manual data entry, and subjective analysis, often leading to delayed issue detection and reactive decision-making. AI-powered real-time project health assessment transforms this process by continuously analyzing key project indicators, detecting risks early, and providing instant insights to prevent delays or budget overruns.

AI-driven project health monitoring tools integrate with task management platforms, financial systems, and team collaboration software to track progress, resource utilization, and potential risks. This enables project managers to make faster, data-driven decisions while keeping teams and stakeholders informed.

How AI Enhances Project Health Assessment
AI-driven tools improve project monitoring by:

- **Providing real-time project status updates** instead of waiting for periodic reports.

- **Detecting early warning signs** of potential risks or delays.

- **Recommending corrective actions** before issues escalate.

A. AI-Powered Real-Time Dashboards

AI consolidates data from multiple sources and presents project health insights on interactive dashboards, ensuring visibility into progress, risks, and team performance.

Example: AI tracks budget utilization, task completion rates, and resource allocation, flagging projects that are behind schedule or over budget.

AI Tools:

- **Tableau AI** – Provides real-time project health analytics.

- **Power BI AI** – Visualizes performance trends and risk indicators.

B. AI-Driven Risk Identification and Alerts

AI continuously scans project data to identify risks related to missed deadlines, scope creep, or resource shortages.

Example: AI detects that 50% of tasks assigned to a critical team are behind schedule, triggering an early warning notification to project managers.

AI Tools:

- **ClickUp AI** – Sends real-time alerts for project risks.

190

- **Microsoft Project AI** – Identifies timeline bottlenecks before they impact delivery.

C. Automated Project Health Scores

AI calculates health scores based on key factors like timeline adherence, budget control, and resource utilization, giving a quick snapshot of overall project stability.

Example: AI assigns a score of 75% to a project based on recent delays and budget overruns, suggesting corrective measures to improve performance.

AI Tools:

- **Monday.com AI** – Generates project health reports with recommended actions.

- **Smartsheet AI** – Predicts future risks based on current project status.

By leveraging AI-powered real-time project health assessments, organizations can proactively manage risks, optimize performance, and ensure project success.

Predictive Performance Analytics

Tracking project performance in real-time is essential, but predicting future performance trends allows project managers to proactively address risks before they impact deadlines, budgets, or resource allocation. Traditional performance analysis relies on historical data and periodic reviews, making it difficult to anticipate potential challenges early enough to prevent delays or cost overruns. AI-driven predictive performance analytics solves this problem by using machine learning, statistical modeling, and historical patterns to forecast project outcomes with high accuracy.

AI can identify potential bottlenecks, resource shortages, budget overruns, and delays before they happen, enabling project managers to make informed, data-driven adjustments. By continuously analyzing real-time project data, AI allows teams to shift from a reactive to a proactive approach, improving efficiency and overall project success.

How AI Enhances Predictive Performance Analytics
AI improves performance tracking by:

- **Analyzing historical project data** to detect performance trends.

- **Predicting potential delays** and recommending corrective actions.

- **Identifying workload imbalances** to optimize resource allocation.

- **Providing early risk assessments** to prevent issues before they arise.

A. AI-Powered Delay and Bottleneck Prediction
AI forecasts potential delays by analyzing past project timelines, team performance, and task dependencies, helping teams mitigate risks before they escalate.

Example: AI predicts that testing will likely be delayed by two weeks based on historical testing durations and current resource availability, allowing project managers to allocate additional testers in advance.

AI Tools:

- **Microsoft Project AI** – Identifies scheduling risks and recommends adjustments.

- **Smartsheet AI** – Forecasts project bottlenecks based on past trends.

B. *Forecasting Resource Utilization and Productivity Trends*

AI analyzes resource workload, efficiency rates, and availability to optimize task assignments and prevent burnout or underutilization.

Example: AI detects that certain team members are consistently overloaded, recommending task redistribution to balance workloads and maintain productivity.

AI Tools:

- **ClickUp AI** – Predicts workload imbalances and suggests redistribution.

- **Monday.com AI** – Reallocates resources dynamically to prevent overloading.

C. *Budget and Cost Overrun Prediction*

AI continuously monitors spending patterns and compares them with historical project data to identify potential budget overruns before they occur.

Example: AI predicts that a project is on track to exceed its budget by 15% and recommends adjustments such as renegotiating vendor contracts or reallocating non-essential expenses.

AI Tools:

- **Tableau AI** – Predicts cost overruns and provides budget insights.

- **Power BI AI** – Analyzes financial trends and suggests cost-saving measures.

By leveraging predictive performance analytics, organizations can anticipate risks, optimize resources, and ensure projects stay on track.

Prompt Techniques for Ongoing Project Evaluation

Continuous project evaluation is essential for ensuring smooth execution, maintaining efficiency, and proactively addressing risks. Traditional evaluation methods rely on manual reporting, periodic reviews, and static KPIs, which can lead to delayed insights and reactive decision-making. AI-powered project evaluation, driven by well-structured prompts, enables real-time monitoring, automated insights, and predictive analytics to help project managers track progress, identify performance gaps, and make data-driven adjustments.

This section provides AI-powered prompt techniques for tracking performance, assessing risks, optimizing resource allocation, and refining project strategies. These prompts can be used with ChatGPT, Microsoft Copilot, Smartsheet AI, and other AI-driven analytics tools to generate actionable insights and ensure continuous project improvement.

AI Prompts for Performance Tracking
AI can provide real-time updates on project progress, helping managers make informed decisions.

A. General Project Status Updates
Prompt Example:

"Provide a summary of the current project status, including completed tasks, ongoing activities, and any potential risks."

AI Output:

- **Completed:** Task A, Task B

- **In Progress:** Task C (50% done), Task D (at risk of delay)

- **Risks:** Resource shortage in Task E

AI Tools:

- **Microsoft Teams AI** – Generates real-time project summaries.

- **Smartsheet AI** – Provides automated project status reports.

B. Identifying Delays and Bottlenecks
Prompt Example:

"Analyze recent project data and identify any tasks that are behind schedule. Suggest strategies to recover lost time."

AI Output:

- **Delayed:** Task F (by 3 days), Task G (at risk of missing deadline)

- **Recommendation:** Reallocate additional team members to accelerate completion.

AI Tools:

- **Monday.com AI** – Detects delays and recommends workload adjustments.

- **ClickUp AI** – Predicts scheduling risks and suggests solutions.

AI Prompts for Risk Assessment
AI can detect early warning signs and recommend strategies for risk mitigation.

A. Predicting Project Risks
Prompt Example:

"Evaluate the project's current risk factors and predict potential issues that could impact delivery timelines or budget."

AI Output:

- **Budget Overrun Risk:** 10% over initial estimates due to unexpected vendor costs.

- **Timeline Risk:** The testing phase may take longer than expected due to limited resources.

- **Recommendation:** Negotiate better vendor pricing and allocate extra testers.

AI Tools:

- **Oracle Primavera AI** – Identifies financial and scheduling risks.

- **Tableau AI** – Analyzes performance data to highlight potential issues.

AI Prompts for Resource Optimization
AI helps teams balance workloads, prevent burnout, and maximize efficiency.

A. Workload Balancing and Resource Allocation
Prompt Example:

"Analyze current resource utilization and suggest adjustments to optimize team workload and prevent burnout."

AI Output:

- **Overloaded Employees:** Developer Team (120% capacity)

- **Underutilized Resources:** UX Design Team (40% capacity)

- **Recommendation:** Shift 20% of development work to the UX team for faster execution.

AI Tools:

- **ClickUp AI** – Rebalances workload dynamically.

- **Microsoft Project AI** – Suggests real-time resource adjustments.

B. Identifying Skill Gaps and Training Needs
Prompt Example:

"Assess team skill sets and identify any gaps that could impact project success. Recommend upskilling strategies."

AI Output:

- **Skill Gap:** Limited expertise in AI automation testing.

- **Recommendation:** Provide targeted training sessions and consider hiring a specialist.

AI Tools:

- **Workday AI** – Identifies training opportunities based on project needs.

- **LinkedIn Learning AI** – Suggests skill development programs.

AI Prompts for Process Optimization and Continuous Improvement

AI can help refine project workflows and improve efficiency over time.

A. Process Bottleneck Analysis
Prompt Example:

"Identify any inefficient processes slowing down project execution and suggest improvements."

AI Output:

- **Inefficiency:** Frequent approval delays for marketing materials.

- **Recommendation:** Implement automated approvals for standard content.

AI Tools:

- **Asana AI** – Analyzes workflow efficiency.

- **Zapier AI** – Automates repetitive processes.

B. End-of-Project Review and Lessons Learned
Prompt Example:

"Summarize key takeaways from [project name], highlighting successes, challenges, and areas for improvement."

AI Output:

- **Success:** Early adoption of automation reduced workload.

- **Challenge:** Budget forecasting was inaccurate due to unexpected costs.

- **Improvement:** Implement AI-driven cost prediction in future projects.

AI Tools:

- **Notion AI** – Organizes lessons learned from project data.

- **Microsoft Viva AI** – Tracks insights from past project performance.

By leveraging AI-powered prompt techniques for ongoing project evaluation, teams can enhance performance tracking, predict risks, optimize resources, and drive continuous improvement

Chapter 11: AI Tools and Technology Stack

AI-powered project management is only as effective as the tools and systems used to implement it. With a growing number of AI-driven platforms, assistants, and automation solutions, choosing the right tools and integrating them into existing workflows can be challenging. Organizations must understand which AI solutions best suit their needs, how to integrate them seamlessly, and how to build a structured workflow that maximizes efficiency and decision-making.

This chapter provides a concise overview of AI project management tools, including those discussed in previous chapters. It explores integration strategies, selecting the right AI assistants, and building a custom AI-powered project workflow. Additionally, it offers practical implementation guides to help organizations adopt AI tools in a structured, results-driven way.

By leveraging the right AI technology stack, businesses can automate repetitive tasks, optimize scheduling, improve decision-making, and enhance collaboration, ultimately leading to more efficient and data-driven project execution.

Overview of AI Project Management Tools

AI-driven project management tools help teams automate tasks, optimize workflows, predict risks, and improve decision-making. These tools integrate machine learning, natural language processing (NLP), and data analytics to enhance efficiency and ensure seamless collaboration. Below is an overview of AI-powered tools used across different aspects of project management.

AI-Powered Task and Workflow Management Tools
ClickUp AI

ClickUp AI automates task creation, prioritization, and progress tracking. It analyzes workloads, detects potential bottlenecks, and suggests task adjustments to maintain productivity.

Asana AI

Asana AI assists in automating task assignments, tracking milestones, and visualizing dependencies. It provides insights into team workload distribution and project timelines.

Monday.com AI

Monday.com AI optimizes project workflows by automating repetitive tasks, identifying inefficiencies, and recommending process improvements based on historical data.

AI for Scheduling and Resource Allocation
Microsoft Project AI

This tool helps predict scheduling conflicts, optimize timelines, and dynamically adjust resources based on workload and priority changes.

Smartsheet AI

Smartsheet AI provides automated scheduling, real-time resource tracking, and predictive analytics to help teams manage dependencies efficiently.

Oracle Primavera AI

Commonly used in construction and large-scale enterprise projects, Oracle Primavera AI helps with risk analysis, scheduling optimization, and critical path adjustments.

AI for Communication and Collaboration
Slack AI

Slack AI enhances team communication by automating responses, summarizing discussions, and analyzing sentiment to improve collaboration.

Microsoft Teams AI

Microsoft Teams AI generates meeting summaries, tracks key action items, and automates project status updates, ensuring efficient communication.

Otter.ai

Otter.ai provides real-time transcription and meeting summarization, ensuring that important discussions are documented and shared accurately.

AI for Budgeting and Financial Management
Tableau AI

Tableau AI visualizes financial trends, detects budget anomalies, and provides predictive cost analysis, helping teams stay within financial constraints.

Power BI AI

Power BI AI integrates with financial data sources to generate automated reports, highlight cost overruns, and suggest budget optimizations.

Anaplan AI

Anaplan AI assists in budget forecasting, cost scenario planning, and financial risk assessment, making it a valuable tool for finance-driven project management.

AI for Risk Management and Performance Tracking
IBM Watson AI

IBM Watson AI analyzes project risks, detects anomalies, and provides predictive insights to help teams mitigate potential issues before they escalate.

RiskLens AI

RiskLens AI specializes in quantifying and prioritizing financial and operational risks, allowing organizations to take a data-driven approach to risk mitigation.

FICO AI

FICO AI is widely used in fraud detection, risk assessment, and financial forecasting, helping teams prevent budget risks and improve financial decision-making.

AI-Powered Reporting and Decision Support
Notion AI

Notion AI assists with automating project documentation, generating reports, and summarizing key takeaways for project teams and stakeholders.

ChatGPT / Microsoft Copilot AI

These AI assistants generate automated project updates, provide data-driven insights, and assist with report generation, saving time on administrative tasks.

Fireflies.ai

Fireflies.ai records transcribes, and extracts action items from meetings, ensuring that no critical information is lost.

By integrating AI-powered project management tools, organizations can enhance productivity, optimize decision-making, and ensure seamless project execution.

Integration Strategies

Implementing AI-powered project management tools is most effective when they seamlessly integrate with existing workflows, systems, and data sources. Without proper integration, teams may struggle with data silos, disconnected processes, and inefficiencies. A well-planned AI integration strategy ensures that project data flows smoothly across platforms, automation enhances productivity, and AI-driven insights are actionable and reliable.

This section outlines key strategies for integrating AI tools into project management environments, ensuring optimal performance and minimal disruption.

Centralized Data Integration

AI tools work best when they have access to unified, real-time project data. Centralizing data integration ensures that AI models can analyze comprehensive information from various sources, leading to accurate predictions, automation, and insights.

Best Practices:

- Connect AI tools to **existing project management platforms** (e.g., Microsoft Project, Jira, Asana).

- Use **API integrations** to sync AI-driven analytics with financial, scheduling, and task-tracking software.

- Store data in **a centralized cloud repository** to ensure accessibility across departments.

Example:

A team using Smartsheet AI for scheduling and Power BI AI for budgeting integrates both tools via API, allowing Smartsheet to pull real-time budget constraints into scheduling decisions.

Automating Workflows with AI
AI can eliminate repetitive tasks, optimize workflows, and improve task prioritization when integrated with automation platforms.

Best Practices:

- Set up **AI-powered automation rules** to handle repetitive tasks like **status reporting, data entry, and approval processes**.

- Use workflow automation tools like **Zapier, Power Automate, or Asana AI** to connect AI capabilities across applications.

- Enable **task dependencies** so AI can dynamically adjust schedules and assign resources based on real-time data.

Example:

A company integrates Jira AI with Slack AI to automatically update teams when a task status changes, reducing the need for manual notifications.

Enhancing Collaboration with AI Integration
AI tools can improve cross-team communication by ensuring that relevant updates and insights reach the right people at the right time.

Best Practices:

- Integrate **AI-driven communication tools** like **Microsoft Teams AI and Notion AI** for automated meeting summaries and documentation.

- Use **AI-powered chatbots** to provide real-time project updates within messaging platforms.

- Sync **AI task assistants** with calendar apps to schedule reminders and meetings based on project timelines.

Example:

A marketing team integrates Fireflies.ai with Microsoft Teams to automatically generate meeting notes and email summaries after strategy discussions.

Ensuring Data Security and Compliance
AI-powered project management tools process large amounts of sensitive project data, making security and compliance critical.

Best Practices:

- Ensure AI tools comply with **GDPR, SOC 2, and other industry regulations**.

- Use **role-based access controls** to limit AI-generated insights to authorized personnel.

- Regularly audit **AI-driven automation and decision-making** to ensure transparency and accuracy.

Example:

A financial services firm using FICO AI for risk analysis ensures all integrations follow strict data security protocols to protect confidential project data.

Choosing Scalable AI Solutions

AI should be flexible and scalable, allowing organizations to expand capabilities as their project management needs evolve.

Best Practices:

- Start with **small AI integrations** and expand based on performance.

- Choose **modular AI platforms** that allow additional functionalities as needed.

- Regularly review AI-generated reports to **refine integrations and improve efficiency**.

Example:

A growing startup begins with AI-powered task automation using Monday.com AI, then gradually integrates predictive analytics for financial forecasting.

By following strategic AI integration approaches, teams can maximize efficiency, collaboration, and automation without disrupting existing workflows.

Selecting the Right AI Assistants

With the growing adoption of AI in project management, organizations must choose the right AI assistants that align with their needs, workflows, and project complexity. AI-powered assistants can help with task automation, decision support, scheduling, communication, and risk assessment, but selecting the wrong tool can lead to inefficiencies and integration challenges.

This section outlines key factors to consider when choosing an AI assistant, ensuring that teams leverage AI capabilities effectively.

Identifying Project Needs and AI Use Cases
Before selecting an AI assistant, organizations should determine which areas of project management need AI-driven improvements.

Questions to Consider:

- Do we need AI for **task automation and workflow management**? (e.g., **ClickUp AI, Monday.com AI**)

- Are we looking for AI to **enhance communication and collaboration**? (e.g., **Slack AI, Microsoft Teams AI**)

- Do we require AI for **financial forecasting and budget tracking**? (e.g., **Power BI AI, Tableau AI**)

- Is AI needed for **risk management and predictive analytics**? (e.g., **IBM Watson AI, RiskLens AI**)

Evaluating AI Capabilities and Features
Different AI assistants offer various functionalities. Teams should assess how well the AI supports real-time insights, automation, and integrations.

Key Features to Look For:

- **Task Automation:** Can the AI handle repetitive processes? (e.g., **Zapier AI, Asana AI**)

- **Predictive Analytics:** Does it offer data-driven forecasting? (e.g., **Oracle Primavera AI, Smartsheet AI**)

- **Natural Language Processing (NLP):** Can it interpret and respond to queries effectively? (e.g., **ChatGPT, Microsoft Copilot AI**)

- **Integration Compatibility:** Does it sync with existing project management tools?

Ensuring Seamless Integration
AI assistants should fit within the organization's existing technology stack without causing workflow disruptions.

Best Practices:

- Choose AI assistants with **strong API support** for seamless integration.

- Prioritize **cloud-based AI solutions** for accessibility and scalability.

- Select AI tools that offer **multi-platform compatibility** (desktop, mobile, and cloud).

Considering Scalability and Cost
AI assistants should be scalable to accommodate project growth and complexity while remaining cost-effective.

Factors to Evaluate:

- **Pricing models:** Subscription-based, pay-per-use, or enterprise licensing.

- **Scalability:** Can the AI handle increasing project demands over time?

- **Support and training:** Availability of customer support and AI learning resources.

By carefully selecting the right AI assistant, organizations can enhance productivity, automate processes, and improve decision-making, ensuring a seamless AI-powered project management experience.

Building a Custom AI Project Management Workflow

Integrating AI into project management requires more than just selecting the right tools—it demands a structured workflow that ensures seamless automation, efficient collaboration, and data-driven decision-making. A custom AI-powered workflow aligns AI capabilities with specific project goals, team dynamics, and business needs, ensuring that AI enhances productivity rather than creating complexity.

This section outlines a step-by-step approach to designing a tailored AI project management workflow that optimizes efficiency and minimizes manual intervention.

Step 1: Define Workflow Objectives
Before implementing AI, project teams must identify key pain points and desired outcomes.

Questions to Consider:

- What **manual processes** need automation? (e.g., status reporting, scheduling, risk assessment)

- Which areas require **AI-driven decision support**? (e.g., budget forecasting, resource allocation)

- How will AI enhance **team collaboration and communication**?

Example: A marketing team struggling with delayed approvals and misaligned timelines implements an AI workflow that automatically tracks project milestones, assigns approvals, and sends real-time notifications.

Step 2: Choose and Integrate AI Tools
Once objectives are clear, teams can select and integrate AI tools that align with their workflow requirements.

Best Practices:

- Use **AI-powered scheduling tools** (e.g., **Microsoft Project AI**) to automate task assignments.

- Leverage **AI-driven collaboration platforms** (e.g., **Slack AI, Microsoft Teams AI**) for real-time updates.

- Implement **predictive analytics tools** (e.g., **Smartsheet AI**) for performance tracking.

Example: A project manager integrates Power BI AI with a budgeting tool to generate real-time financial reports, reducing manual financial analysis.

Step 3: Automate Workflow Processes
AI can streamline workflows by reducing repetitive tasks and dynamically adjusting schedules, reports, and resource allocation.

Implementation Steps:

- **Set AI triggers** for status updates, risk alerts, and task completions.

- **Automate reporting** using tools like **Tableau AI** to track project KPIs.

- **Use AI assistants** (e.g., **ChatGPT, Notion AI**) to generate meeting summaries and documentation.

Example: AI automatically flags delays, reallocates resources, and notifies teams, preventing workflow disruptions.

By designing a custom AI-powered project management workflow, teams can eliminate inefficiencies, enhance collaboration, and drive continuous improvement.

Practical Implementation Guides

Successfully adopting AI in project management requires a structured approach to implementation, integration, and optimization. Without a clear plan, AI adoption can lead to disconnected processes, underutilized tools, and resistance from teams. This section provides a step-by-step guide to implementing AI-powered project management solutions effectively.

Step 1: Start with a Pilot Program
Before full-scale implementation, organizations should test AI tools in a controlled environment to evaluate their effectiveness.

Steps:

- Select a small project or department for the initial AI rollout.

- Implement one or two AI tools (e.g., AI-powered scheduling or reporting tools) to measure impact.

- Collect feedback from team members to refine AI usage.

Example: A software development team tests Jira AI for backlog prioritization before expanding it to the entire company.

Step 2: Train Teams and Establish AI Best Practices
To ensure successful adoption, teams must understand how to interact with AI-driven tools.

Best Practices:

- Provide hands-on training for AI-powered platforms like Microsoft Copilot or ClickUp AI.

- Establish AI usage guidelines to ensure consistency.

- Encourage teams to use AI prompts effectively for communication and reporting.

Step 3: Monitor Performance and Optimize AI Usage
AI should be continuously monitored and refined to align with evolving project needs.

Optimization Steps:

- Regularly review AI-generated insights for accuracy.

- Adjust AI parameters based on team feedback and project complexity.

- Scale AI adoption by integrating additional tools as workflows mature.

Example: A finance team using Power BI AI for budgeting adjusts AI settings to improve forecasting accuracy.

By following these practical implementation steps, organizations can seamlessly integrate AI, enhance efficiency, and drive long-term project success.

Chapter 12: Case Studies and Real-World Applications

The integration of Artificial Intelligence (AI) into project management has revolutionized various industries by enhancing efficiency, accuracy, and decision-making processes. This chapter delves into real-world applications of AI in project management, highlighting detailed case studies from diverse sectors. These examples illustrate the tangible benefits and challenges encountered during AI implementation, offering valuable insights for organizations considering similar advancements.

Case Study 1: Amarra's Integration of AI in E-commerce Operations

Company Overview

Amarra, based in New Jersey, is a global distributor of special-occasion gowns, catering to a diverse clientele seeking high-quality formal wear. Recognizing the competitive nature of the fashion industry and the growing importance of e-commerce, Amarra sought innovative solutions to enhance operational efficiency and customer satisfaction.

Challenges Faced

Prior to AI integration, Amarra faced several challenges:

- **Content Creation:** Crafting engaging and accurate product descriptions was time-consuming, leading to delays in product listings.

- **Inventory Management:** Overstocking issues resulted in increased storage costs and potential losses due to unsold inventory.

- **Customer Service:** Handling a high volume of customer inquiries manually led to longer response times and strained resources.

AI Implementation

In 2020, Amarra embarked on a strategic initiative to integrate AI into its operations[11]:

- **Product Descriptions:** Utilizing ChatGPT, an AI language model, Amarra automated the creation of product descriptions. This tool generated engaging and accurate content, reducing the time required for content creation by 60%.

- **Inventory Management:** Amarra implemented an AI-powered inventory management system that analyzed sales data, trends, and customer preferences. This system optimized stock levels, leading to a 40% reduction in overstock.

- **Customer Service:** AI-driven chatbots were deployed to handle routine customer inquiries, addressing 70% of queries without human intervention. This allowed human agents to focus on more complex issues, improving overall customer satisfaction.

Outcomes and Benefits

The integration of AI yielded significant improvements:

- **Efficiency:** Automating content creation and customer service tasks led to faster operations and reduced staff workload.

---

[11] https://www.businessinsider.com/wholesale-formal-gown-distributor-using-ai-for-ecommerce-operations

- **Cost Savings:** Optimized inventory management reduced storage costs associated with overstocked items.

- **Customer Experience:** Quicker response times and accurate product information enhanced the overall shopping experience, leading to increased customer loyalty.

Challenges and Solutions

Despite the benefits, Amarra encountered challenges:

- **Balancing Automation with Human Touch:** Ensuring that automation did not compromise personalized customer interactions required continuous monitoring and adjustments.

- **System Integration:** Integrating AI tools with existing systems posed technical challenges, necessitating collaboration between IT and operational teams.

- **Managing AI Bias:** Addressing biases in AI models was crucial to maintain fairness and accuracy, leading to ongoing model training and evaluation.

Key Takeaways

Amarra's experience offers valuable lessons:

- **Strategic Implementation:** Identifying specific areas where AI could have the most impact ensured targeted improvements.

- **Continuous Monitoring:** Regularly evaluating AI performance and making necessary adjustments to maintain system effectiveness.

- **Employee Involvement:** Engaging staff in the AI integration process facilitated smoother transitions and acceptance of new technologies.

Conclusion

Amarra's successful integration of AI into its e-commerce operations demonstrates the potential of AI to transform business processes. By addressing specific operational challenges through targeted AI applications, Amarra enhanced efficiency, reduced costs, and improved customer satisfaction, positioning itself competitively in the fashion industry.

Case Study 2: BHP's AI-Driven Optimization in Mining Operations

Company Overview

BHP Group, headquartered in Melbourne, Australia, is one of the world's leading resources companies, specializing in the extraction and processing of minerals, oil, and gas. With operations spanning across continents, BHP is committed to leveraging technological advancements to enhance productivity, safety, and sustainability in its mining operations.

Challenges Faced

The mining industry presents unique challenges that can impact operational efficiency and profitability:

- **Complex Data Management:** Mining operations generate vast amounts of data from various sources, making it challenging to process and analyze information effectively.

- **Operational Efficiency:** Optimizing equipment performance and processing operations is critical to maximize yield and reduce costs.

- **Safety Concerns:** Ensuring the safety of personnel in hazardous environments requires continuous monitoring and proactive measures.

AI Implementation

BHP recognized the potential of Artificial Intelligence (AI) to address these challenges and embarked on a comprehensive AI integration strategy:

- **AI Platform Development:** In 2022, BHP collaborated with AI architect Julian Adam Wise[12] to develop an AI platform tailored for its Latin American operations. This platform facilitated the seamless integration of machine learning models into existing systems, enhancing data processing capabilities.

- **Predictive Maintenance:** By deploying AI algorithms to analyze equipment data, BHP implemented predictive maintenance schedules, reducing unexpected machinery failures and downtime.

- **Process Optimization:** AI models were utilized to analyze mineralogy data and processing parameters[13], leading to optimized recovery rates in mineral processing plants.

---

[12] https://en.wikipedia.org/wiki/Julian_Adam_Wise
[13] https://www.researchgate.net/publication/374625102_Improving_BH Pproject_delivery_timelines_with_Machine_Learning_Operations_and _Cloud_Technology

- **Safety Monitoring:** AI-driven sensors and monitoring systems were installed to detect potential safety hazards in real time, enabling prompt responses to emerging risks.

Outcomes and Benefits

The integration of AI into BHP's mining operations yielded significant improvements:

- **Enhanced Productivity:** AI-driven process optimizations led to increased throughput and higher mineral recovery rates.

- **Cost Reduction:** Predictive maintenance minimized unplanned downtime, resulting in substantial cost savings associated with equipment repairs and production losses.

- **Improved Safety:** Real-time monitoring and predictive analytics contributed to a safer working environment, reducing the incidence of accidents and enhancing compliance with safety regulations.

Challenges and Solutions

Despite the benefits, BHP encountered challenges during AI implementation:

- **Data Integration:** Combining data from disparate sources required the development of robust data governance frameworks to ensure data quality and consistency.

- **Change Management:** Introducing AI technologies necessitated cultural shifts within the organization. BHP addressed this by engaging stakeholders through training programs and demonstrating the value of AI in enhancing operational outcomes.

- **Technical Expertise:** Building and maintaining AI models required specialized skills. BHP invested in upskilling its workforce and collaborating with external experts to bridge knowledge gaps.

Key Takeaways

BHP's experience offers valuable insights for organizations considering AI integration:

- **Strategic Alignment:** Aligning AI initiatives with business objectives ensures that technology investments drive meaningful outcomes.

- **Collaborative Approach:** Engaging cross-functional teams fosters a culture of innovation and facilitates the successful adoption of new technologies.

- **Continuous Improvement:** Regularly evaluating AI performance and incorporating feedback loops enables organizations to adapt to evolving operational needs.

Conclusion

BHP's strategic integration of AI into its mining operations demonstrates the transformative potential of technology in traditional industries. By addressing specific operational challenges through targeted AI applications, BHP enhanced productivity, reduced costs, and improved safety, solidifying its position as a leader in the resources sector.

Amarra and BHP's journey underscores the importance of embracing technological innovation to drive operational excellence. The lessons learned from their AI integration efforts provide a roadmap for other organizations seeking to harness the power of AI in different environments.

Chapter 13: Future of AI in Project Management

The integration of Artificial Intelligence (AI) into project management is transforming how organizations plan, execute, and monitor projects. AI's ability to process vast amounts of data, predict outcomes, and automate routine tasks offers unprecedented opportunities to enhance efficiency and decision-making. This chapter explores emerging AI technologies, predicted industry trends, strategies for preparing for AI-driven project management, and the ethical considerations inherent in human-AI collaboration.

Emerging AI Technologies

The landscape of AI in project management is continually evolving, with several technologies poised to make significant impacts:

1. AI-Powered Project Scoping Tools

Tools like Tara AI utilize machine learning to predict technical tasks, timelines, and resource allocations for new software projects, streamlining the project initiation phase.

2. Intelligent Project Management Platforms

Platforms such as Easy Redmine integrate AI to enhance various aspects of project management, including risk assessment, resource management, and task prioritization.

3. AI Agents and Assistants

The development of AI agents capable of automating complex tasks is gaining momentum. Companies like

Johnson & Johnson and eBay are leveraging AI agents to enhance operations in areas such as drug discovery and content generation.

4. Predictive Analytics Tools

AI-driven predictive analytics tools analyze historical data to forecast project outcomes, identify potential risks, and suggest corrective actions, enabling proactive project management.

5. Natural Language Processing (NLP) Applications

NLP technologies facilitate the analysis of unstructured data, such as emails and meeting notes, extracting valuable insights that inform decision-making and improve communication.

Predicted Industry Trends

The future of AI in project management is characterized by several key trends:

1. Ubiquity of AI Tools

AI tools are becoming integral to project management, assisting in task automation, data analysis, and decision support, thereby enhancing productivity.

2. Emergence of AI-Savvy Managers

Managers proficient in AI technologies are emerging as valuable assets, capable of effectively integrating AI into project workflows to optimize performance.

3. Enhanced Collaboration Between Humans and AI

The synergy between human intuition and AI's analytical capabilities is leading to more effective project management strategies, combining creativity with data-driven insights.

4. Shift Towards Proactive Risk Management

AI's predictive capabilities enable the anticipation of potential project risks, allowing for the implementation of mitigation strategies before issues arise.

5. Customization and Personalization of AI Solutions

AI solutions are increasingly tailored to specific organizational needs, offering customized functionalities that align with unique project requirements and industry contexts.

Preparing for AI-Driven Project Management

To effectively embrace AI in project management, organizations should consider the following strategies:

1. Invest in AI Education and Training

Providing team members with education and training on AI technologies ensures they possess the necessary skills to leverage AI tools effectively.

2. Foster a Culture of Innovation

Encouraging a culture that embraces technological advancements and experimentation with AI fosters adaptability and continuous improvement.

3. Develop a Clear AI Integration Strategy

Establishing a comprehensive strategy for AI integration ensures alignment with organizational goals and maximizes the benefits of AI adoption.

4. Collaborate with AI Experts

Partnering with AI specialists or consultants provides valuable insights and guidance, facilitating a smoother transition to AI-driven project management.

5. Pilot AI Initiatives

Implementing pilot programs allows organizations to assess the impact of AI on specific projects, gather feedback, and refine AI strategies before full-scale deployment.

Ethical Considerations and Human-AI Collaboration

As AI becomes more prevalent in project management, addressing ethical considerations and fostering effective human-AI collaboration are paramount:

1. Addressing Bias in AI Algorithms

Ensuring AI algorithms are free from biases is crucial to maintaining fairness and equity in project management decisions. Regular audits and diverse data sets can mitigate bias.

2. Ensuring Transparency and Explainability

AI systems should operate transparently, with decision-making processes that are understandable to users, fostering trust and accountability.

3. Maintaining Data Privacy and Security

Protecting sensitive project data is essential. Implementing robust data security measures safeguards against breaches and unauthorized access.

4. Defining the Scope of AI Autonomy

Clearly delineating the tasks AI systems can perform autonomously versus those requiring human oversight ensures appropriate levels of control and responsibility.

5. Promoting Ethical AI Use

Establishing guidelines for the ethical use of AI in project management prevents misuse and aligns AI applications with organizational values and societal norms.

In conclusion, the future of AI in project management holds immense potential to transform how projects are conceived, executed, and evaluated. By staying abreast of emerging technologies, anticipating industry trends, preparing strategically for AI integration, and addressing ethical considerations, organizations can harness AI's capabilities to achieve greater efficiency, innovation, and success in their project endeavors.

Additional Resources

This book has been published as a resource for you, and we hope you enjoyed it.

If you have any feedback or suggestions, please feel free to send us an email at lynn@talentcoreservices.com.

You can also download additional resources and find our other products at www.talentcoreservices.com.

Also published by this author:

Life's a Project: Discover How Project Management Principles Can Revolutionize Your Life!

Life's a Sprint: Applying Scrum and Agile Principles for Daily Success

Thank You

www.ingramcontent.com/pod-product-compliance
Lightning Source LLC
Chambersburg PA
CBHW071554210326
41597CB00019B/3241